HARD AGROUND

HARD AGROUND

The Story of the Argo Merchant Oil Spill

Ron Winslow

W · W · NORTON & COMPANY · INC · *New York*

Copyright © 1978 by Ron Winslow
Published simultaneously in Canada by George J. McLeod Limited,
Toronto. Printed in the United States of America.
All Rights Reserved
First Edition

Library of Congress Cataloging in Publication Data
Winslow, Ron.
   Hard aground.
     1. Oil spills—Massachusetts.   2. Shipwrecks—Massachusetts.
3. Oil pollution of the sea—Massachusetts.
4. Argo Merchant (Ship)   I. Title.
GC1212.M4W56 1978   974.4'9   78–8121
ISBN 0-393-05687-2

Book design by A. C. Goodman
Typefaces are Video Gael and Folio Bold Extended
Manufactured by The Haddon Craftsmen

1 2 3 4 5 6 7 8 9 0

for DON MURRAY

Contents

ACKNOWLEDGMENTS 9

1 The Grounding 13
2 The Search 26
3 The Response 38
4 The Damage Control Team 51
5 The Blackout 65
6 The Condition 77
7 The Rigging 93
8 The Progress 104
9 The Evacuation 116
10 The Strike Team 132
11 The Plan 163
12 The Hawsepipe 177
13 The Current 190
14 The Breakup 200
15 The Ship 210
16 The Last Voyage 229
17 The Slick 243
18 The Aftermath 262

EPILOGUE 277

Acknowledgments

This book is an outgrowth of my reporting on the *Argo Merchant* incident in December 1976 and January 1977 for the *Providence Journal.* Chuck Hauser and Jim Wyman, two editors at the newspaper, agreed to give me leaves of absence in the spring and fall of 1977, without which I could not have completed the project. I owe special thanks to Al Johnson, my city editor, for his flexibility and understanding while I finished the book after returning to work. I am also indebted to Joel Rawson, another editor at the *Journal,* for his efforts to encourage an in-depth narrative approach to news writing, a form foreign to most newspaper reporters.

Several people in the Coast Guard gave up hours of their time to grant interviews. I especially appreciate the cooperation of the Atlantic Strike Team, the Marine Safety Office in Boston, the Coast Guard Air Station on Cape Cod, the commanding officers of the *Vigilant* and the *Sherman,* and the public affairs office of the First Coast Guard District.

Several of my friends reviewed portions of the manuscript and made valuable suggestions. They include Candy Page, Bruce Butterfield, Ed Brodeur, Joel Rawson, Judy Stark, Lou Ureneck, Bernie Sullivan, and Bob Stickler. Dorothy Meegan typed the final manuscript.

Outside of my family, I am especially grateful to four other people: Starling Lawrence, my editor at Norton, for his perceptive suggestions and for his faith in and patience with a new writer; Jan Harayda, who offered constant encouragement and helpful suggestions from the time the book was only an idea through its completion; Wayne Worcester, who read nearly every word in early drafts, who made several key suggestions, particularly regarding organization of the material, and whose contributions show up in nearly every chapter; and Don Murray, my writing teacher at the University of New Hampshire, for whom the dedication is a small token of my appreciation for his inspiration and guidance going back to 1968 when I was fortunate enough to walk into his classroom.

The contributions of my family have been overwhelming. My wife's mother, Arline Chapman, my mother, Doris, and my sister Sandy all labored for hours with a tape recorder and typewriter, transcribing some of my interviews. Along with my wife's father, Lloyd, and my father, Ron, they also read the book as it progressed. By even more than their tangible contributions, I especially appreciate my parents' genuine interest and excitement in the endeavors of their son.

My wife, Carol, has endured the project with patience, love, and grace. She has sacrificed without complaint; she had read and reread segments of the book with an editor's eye; she has answered my endless questions and comments on the research and writing; she has supported the book wholeheartedly from the start. In many ways, it is as much hers as it is mine.

HARD AGROUND

1

The Grounding

The telephone rang in the captain's quarters at 12:45 A.M. jolting Georgios Papadopoulos awake. The forty-three-year-old master of the *Argo Merchant* had expected the call, but still, coming in the middle of the night, it startled him for an instant as he reached for the telephone near his bed.

"Captain, this is Dedrinos on the bridge," the voice crackled. "It is nearly oh-one-hundred. We are approaching the light."

Papadopoulos, a short man with a black mustache as thick as wire on his long, pointed face, awakened quickly. He was accustomed to having his sleep interrupted at sea since he always came to the bridge when his ship was to make a major course change or approach a port or a light station. He told Georgios Dedrinos, the second officer, he would be right up.

It was December 15, 1976, and the *Argo Merchant*, a 29,870-ton Liberian-registered oil tanker, was plodding through increasingly heavy seas in the North Atlantic Ocean south of Nantucket Island, Massachusetts. Her hull measured six hundred forty-one feet, longer than two football fields. She was bound for the once famous whaling port of Salem, thirty miles north of Boston, where she was scheduled late that day to discharge 7,677,684 gallons of heavy industrial fuel oil from her white, rust-stained hull. It was enough oil to supply 18,000 New

England homes with electricity for a year.

The weight of the cargo, more than 27,500 tons, pulled the hull into the water to a draft of thirty-five feet, so deep that the ship resembled an iceberg; nearly eighty percent of her bulk was beneath the surface as she moved through the water.

The tanker was approaching the Nantucket Lightship, one of the most important navigation sentries in the world. Moored forty-eight miles southeast of Nantucket, at the crossroads of two major shipping lanes, its shining beacon warns mariners of the treacherous and legendary ship's graveyard to its north and west known as Nantucket Shoals.

Dedrinos had begun his watch at midnight and, with the officer of the preceding watch, had estimated the tanker's position as thirty-six miles southwest of the lightship. Though it was just a guess, known as a dead reckoning, Dedrinos was happy with the position because it confirmed an earlier estimate that the tanker would pass just to the southeast of the navigation marker between 3:30 and 4 A.M. He expected to see it about an hour before then.

Papadopoulos came up on the bridge precisely at 1 A.M. and went with Dedrinos into the chartroom. Laid out on a table was a chart showing the waters off the North Atlantic Coast, from Cape Hatteras, where the ship had set its current course two days before, past Nantucket to Cape Sable, Nova Scotia. Using a ruler, they lined up their noon positions of the previous days as they had been plotted on the chart and projected their course northward. They looked at each other quickly and lined it up again.

"It goes right up on the shoals, to the port of the lightship," Dedrinos remarked, without a hint of concern in his voice.

The captain nodded. He was not alarmed. Staring calmly at the chart for a moment, he reviewed the course once again. The wind and seas were coming from the northwest, as they had been for much of the past forty-eight hours, and would tend to push the vessel to starboard, to the outside of the light, he thought. In fact, he had purposely adjusted his course a few degrees to the north two days before to compensate for those effects, and the ship was on a bearing that pointed directly at the shallow, rocky water of the shoals. But Papadopoulos was certain that the ship had drifted to the east as it headed northeast during the last one hundred miles and was on an actual course that would pass safely to the east of the lightship and the shoals.

Dedrinos pulled out another chart, which showed the direction of natural ocean currents in that area during November. No December chart was on board, so he picked the closest one on hand.

"The currents are offshore," he reported to the captain. "They will carry us to starboard."

"Yes, you're right, I know it," Papadopoulos replied.

What the captain did not know was that the December chart would have shown currents pushing in the opposite direction. If they were having any effect at all, they were pushing the vessel slowly to the west of the light.

He leaned over the table, checking and rechecking the chart on a routine approach to the carefully marked waters and the lightship ahead.

He looked up from the chart at Dedrinos, who, at twenty-seven, was serving on his first ship as an officer.

"We will keep to our present course," the captain said.

Papadopoulos left the chartroom and walked out on the starboard lookout, just off the bridge. He felt the rush of cold December air on his face. Just below him, the bow was rising and falling sluggishly as it surged through the swells, taking a pounding from the seas against her port side. Occasionally, a wave broke over the deck and the captain could hear the wind-whipped spray shower the deck's huge cargo pipes with a salty mist.

He peered out over the seas in the likely direction of the lightship. Except for the dim glow of the tanker's running lights and the reflection of the moon off the water, it was black. He saw nothing. But at 1:30 he didn't expect to see it. The beacon beams its light from a perch fifty-five feet above the water. Its range, according to Papadopoulos' chart, was fourteen miles. The captain figured he was about twenty-five miles away. And the visibility was only seven miles that night, so it was unlikely anyone would see the light much before three.

The captain did not cut an imposing figure. He was short, about five feet, eight inches tall and so thin his clothes seemed to hang on him. His black but graying hair was receding at the temples, and his complexion was dark and normal for a man of forty-three, not ruddy or weatherbeaten. Only in his eyes, which were set deep under bushy eyebrows and tired and wrinkled in the corners from long hours of squinting over the waters from the bridge, did he show the wear of his profession. But his appearance was no reflection on his record as a seaman. During much of his twenty-five-year career as a master, he had skippered tankers owned by Aristotle Onassis, the late Greek shipping magnate. And except for a couple of minor oil spills, an affliction common to tankers, he had never had an accident at sea.

When he came off the lookout, Papadopoulos checked the radar. Pressing his face against the rubber eye shield, he peered into the scope for several minutes, watching the relentless sweep of the thin beam of light as it searched for objects on the water. Except for a few blips off to his port side, the screen was blank. Fishing boats, the captain figured. He checked the settings and adjusted the radar up and down its scale from twelve to twenty-four to forty-eight miles. Still, except for the fishing boats, nothing. No stationary spot on the screen that could be the lightship. The captain was surprised. Perhaps the weather had slowed the vessel down and they were behind schedule, he thought.

Out on the lookout, Dedrinos took his turn scanning the blackness as the moonlight flickered on the waters, playing tricks with his vision. More than once he mistook the moonshine for the light of a fishing boat. And shortly after 1:30 he did see a group of boats off to port, the same ones the captain had spotted on the radar. But he knew he had not seen the light. There would be no confusing its bright beacon with the dots from the fishing boats and the reflections from the moon.

Papadopoulos decided to try the radio direction finder, commonly known as the RDF. In addition to its light, the Nantucket Lightship sends out a radio signal, with a range of one hundred miles, that ships equipped with an RDF can pick up to determine their bearing on the light. The Nantucket signal is three dashes followed by a single dash—Morse code for O-T.

The captain clamped the earphones over his head and started working the dials that regulated the volume and turned the antenna on top of the bridge. He heard nothing. For fifteen minutes he turned the dials, and for fifteen minutes he got nothing but silence. Frustrated, he tore the earphones from his head.

"I don't get anything here, nothing at all," he told Dedrinos, who had come in from the lookout. "There is nothing on the radar either."

"All I've seen is fishing boats," said the second officer. "But it is early. We wouldn't see it yet."

"We must be behind. The weather must have held us back."

Dedrinos nodded and tried to look relaxed. But he was uneasy. He could not understand why neither the radar nor the RDF had picked up the lightship. But he did not mention his concern to the captain.

For the next two hours, Papadopoulos and Dedrinos crisscrossed the bridge, from the lookouts to the radar to the chartroom, searching more urgently at each position than the last. The captain rechecked the RDF every fifteen minutes. Except for the moonlight and the boats, the sea remained black. The radar shows nothing that could possibly be the lightship. The earphones on the RDF remained silent. A knot started to grow in Dedrinos' gut. When he came on the bridge at midnight, he had expected to be alongside the lightship by the end of his watch at 4 A.M. Now it was nearly four and he had not even seen it.

Dedrinos was typical of the three deck officers aboard the *Argo Merchant*. His boyish face looked closer to eighteen than twenty-seven and he was eager for a career at sea. He had gotten his second mate's license only three months ago, and his head was filled with book knowledge of magnetic compasses, radar sets, and ocean currents—knowledge unseasoned by the wind and the sun and long hours on the bridge. And if there was a brashness or a cockiness to the authority he had in his youth, it was suppressed out of respect when the captain was on the bridge. Though Dedrinos was officially in charge of navigating the vessel during his watch—making sure the helmsman kept the vessel on course, checking the instruments and keeping the charts in order—he yielded his authority when the captain was with him. After all, Dedrinos felt, the captain is in charge of the vessel. He knows what course to choose in charting a voyage and when to deviate from it. Except for the anxiety in his face, Dedrinos kept to himself his doubts about Papadopoulos' insistence on maintaining the present course. But he decided that although his watch was nearly over, he would remain on the bridge until the light was in sight.

At 3:50, Georgios Ypsilantis, the thirty-three-year-old chief mate, climbed out of bed, dressed quickly, and headed for his turn on watch. Before checking with Dedrinos and Papadopoulos, who were inside the glass-paneled bridge, he went out on the starboard lookout to accustom his eyes to the dark and to freshen his sleepy face in the cold, brisk air.

Shortly before 4 A.M., he came in from the lookout. He checked the compass, the helm, and noticed the radar and the depth finder were operating.

"What happened, did you find the light?" he asked in a tone that anticipated an affirmative answer.

"No," the captain replied. "We haven't seen anything."

Ypsilantis was stunned. He looked at the captain, then at Dedrinos. Their faces were stern and drawn and plainly worried. He had never seen them so nervous. The air around him grew taut.

"The lightship, you must have seen the lightship. You should have seen it an hour ago. The light should be alongside right now."

Dedrinos shook his head. "That's what we thought. We haven't seen a thing."

"We must be behind on our course," the captain said.

The chief officer walked quickly back to the starboard lookout. His eyes scanned the horizon. He looked to port. To starboard. Dead ahead. He saw nothing. He stepped back inside.

"Now captain, we should have seen that lightship," he said again, his voice growing sharp.

He went to check the charts. As he studied the ship's course, he wondered aloud whether the heading should be changed a few degrees to the east.

"No. We'll let it go," Papadopoulos said. "We've had the currents and the seas to port, and we are a little behind. The course is okay."

Ypsilantis rarely questioned the captain's judgment. Like Dedrinos, the chief officer relied on the captain to determine the course. And he had never had a reason before to challenge him. But this time, he pressed further.

"Look captain, something has to be done. The course is inside the light. We might crash on the rocks and that would be it. We might drown."

"We are in no danger." The captain's voice was impatient. "The fathometer is operating. We are in safe water. We are keeping to this course."

"Okay, captain."

Ypsilantis turned away. He looked again at the fathometer, an electronic depth finder. It was working and the depth was adequate. But that did little to calm him. They should have seen the lightship. He went over to the radar. The sweeping light picked up nothing. For ten minutes he worked at it, adjusting it to different scales. Still, he had no image on the screen.

While Dedrinos returned to the lookout, the captain tried the RDF again. He made a special effort to pick up not only the Nantucket O-T but also the signals from other radio beacons in the area which operate in sequence with the lightship on the same channel. Silence. His confidence in the course he had so insistently kept to for the past three hours began to falter. For the first time, he wondered to himself whether he should change the course. But he didn't consider it for long.

At 4:30, Ypsilantis got his first chance at the RDF. He put the earphones on and twisted dials back and forth. Within a minute, he got a signal. Dah-dah-dah, dah. He turned the dials again and listened carefully. Dah-dah-dah, dah.

"Captain, I've got it! I've got the O-T!"

Papadopoulos nearly sprinted back to the RDF. He took the earphones and listened.

"That's it, you have it! You do have it." For the first time in three hours, Papadopoulos smiled broadly. Ypsilantis could see tension drain out of the muscles in the captain's face.

The chief officer took the earphones back. As he turned the dial slightly, the pitch of the signal grew loud, then soft, then loud again. He stopped the dial at the point where he could barely hear the O-T, then took the bearing. It was nearly dead ahead.

"The lightship is right off the bow, captain. We're headed right for it."

Ypsilantis twisted the headset and put one earphone to his ear. He gave the other one to the captain so they could both listen at the same time. They nearly bumped their heads together.

The dead-ahead bearing indicated that they had not yet passed the lightship. It confirmed Papadopoulos' hunch that the vessel was behind. And it renewed his confidence in the course. Changing it was out of the question now. Both he and the chief officer had found the bit of good news they were looking for and they latched on to it. They felt they were not in danger.

But over the next half hour, none of the other indicators fell in line. The radar screen was still blank. The sea was still black. Not even the loom of the light was visible on the horizon. And the depth finder was showing increasingly shallower water beneath them. By 5 A.M., the captain noticed readings of twenty fathoms, about one hundred twenty feet. He did not think to check the measurement with the depth readings on his chart.

Ypsilantis went out on the lookout, hoping to see any kind of light. A boat. A ship. Something they could reach by radio to check their course and position. But the fishing vessels that had seemed to dot the waters earlier were gone. The *Argo Merchant* was alone and lost on a vast sea of blackness.

By 5:30, Ypsilantis was desperate. He decided to take an astral fix, an estimate of the vessel's position based on the angle of stars above the horizon. He knew it was too early. To get an accurate reading, he needed to see the stars and the horizon at the same time, a condition that exists only at twilight and just before dawn. He had planned to take a fix between 6:05 and 6:15. But he couldn't wait. Although the horizon was too dark for a good fix, the sky was brightening faintly in the east. He could at least come close, he thought. But he wished the vessel had a loran set. He could have pinpointed the position electronically, without the stars.

He peered carefully through a sextant, squinting and straining his eyes in an effort to discern the almost invisible line where the black sea and the black sky met. He checked and rechecked the measurements until he was reasonably sure they were correct. Using special navigation tables, he converted his measurements to other numbers and added them up to arrive at the coordinates. At 5:45, he plotted the position on the chart. They were forty miles southeast of the lightship, on a course that would take them to the inside of the marker, toward the shoals.

Neither the captain nor the chief officer were satisfied with the fix. While it reconfirmed the captain's belief that the weather had slowed the ship down, the position indicated they had not even reached the dead reckoning Dedrinos had plotted nearly six hours before. They couldn't be that far behind schedule. Still, they figured the fix was within fifteen miles of accuracy. And everything within that range was safe water.

Ypsilantis was still concerned, however, that the tanker was headed inside the light. He thought the captain would be wise to change the course slightly to the right. But he said nothing. The captain could see the charts as plainly as he could, the chief officer thought. It was Papadopoulos' business to determine the course. Besides, the pressure was off. While the fix was hardly satisfactory, it surely indicated, he felt, that they could wait another half hour. Then conditions would be right and he would get a good measurement from the horizon. The ordeal would be over. If necessary, the course could be corrected then.

But under the pressure of the moment, the chief officer in making his calculations had added wrong. His fix was forty miles off. Had he added correctly, he would have come up with coordinates showing that the *Argo Merchant* had already passed to the inside of the Nantucket Lightship and was within fifteen miles of shoal water that was only thirty feet deep.

Five minutes later, Papadopoulos failed to heed the only warning that could have compensated for Ypsilantis' error. The depth finder read fifteen fathoms, ninety feet. The captain raised his eyebrows in disbelief. The chart showed much deeper water in the area around their 5:30 position. A fifteen-fathom reading in the vicinity of the lightship is a clear warning that a vessel is dangerously near the shoals. But the captain didn't trust it. His RDF continued to pick up the lightship's radio signal. The dead ahead bearing had not wavered more than a degree since they had first heard it ninety minutes before. He joined Ypsilantis and Dedrinos, who were out on the starboard lookout, searching the ocean, waiting out the last eight minutes before the chief officer could take another fix.

The captain detected the danger first. He heard a subtle but undeniable change in the sound of the wave, a sign that no electronic instruments would pick up, only an experienced seaman. It was the sound of shallow water. The captain had no time to react. The ship shook lightly, as if a tremor were spreading through it. There was no violent crash. No piercing noise. The officers hardly had to move to keep their balance. The helmsman thought they had plunged into a huge wave. Then they felt a bump.

"Hard aport!" Ypsilantis hollered.

The helmsman put the wheel hard to the left. The bow turned slightly, but the ship's belly was skidding on the bottom. The tanker shuddered to a halt. Papadopoulos, Ypsilantis, and Dedrinos looked at each other. Their wide-eyed faces turned white. Instinctively, all three looked at their watches. It was 6 A.M. From the lookout, they could hear huge crashing waves that sounded like breakers rolling ashore. They had no idea where they were.

Papadopoulos ordered the engines to a dead stop. Then he pushed the emergency alarm. The sound of a siren screeched through the ship. Crewmen bolted upright in their bunks, as if a bomb had gone off beneath them. Engineering officers scampered down narrow walkways to the engine room and waited anxiously for information and instructions. Other crew members grabbed life jackets and headed for the bridge or to lifeboats. No one knew what had happened.

On the bridge, Ypsilantis agreed to go to the main deck and take soundings around the ship, to see how deep the water was and to determine if they could get her off. He told the captain not to do anything until he got back.

As he left the bridge, Ypsilantis could see waves crashing over the deck. He could feel the sway of the ship as it rocked back and forth. Getting soundings would be difficult, he thought.

The captain waited impatiently. He wanted to get the ship off. He was scared. Seven minutes. Ten minutes. Twelve minutes. Ypsilantis had not returned.

"Full astern!" he shouted, pushing a lever that telegraphed the order to the engine room.

The captain had slammed his vessel into reverse. The ship groaned at first, then rumbled as the propeller slowly increased speed and strained against the weight of the oil-laden hull. Thick black smoke belched from the stack. The entire tanker shook in a violent vibration. In the engine room, crew members stumbled and fell. Others grabbed for handrails, ladders, anything they could reach to keep their balance.

Suddenly a pipe connected to one of the tanker's two boilers ruptured. Then a valve that regulated the intake of cooling sea water into the boilers sheared off. The ocean gushed into the engine room.

The officers and crew searched frantically for the source of the leak. They pulled switches, closed valves and inspected pipes they could see. But water was swirling around them. They couldn't see the floor. They could hardly stand up. And they could not stop the flow of water into the ship.

When he felt the vibration of the vessel, Ypsilantis was furious. He dashed back to the bridge. The tension and frustration that had been building for the past two hours during the futile search for the lightship had reached its threshold. When he got back to the captain, it erupted.

"Didn't I tell you to wait?" he hollered. "What the hell are you doing? You were going to wait until I had the soundings. You've lost the ship, goddammit. You lost her!"

"We are stuck aground!" the captain shouted back. "We've got to get her off. I can't wait all morning for the soundings."

"Well, you should have waited. The ship is gone."

Ypsilantis was sure he felt the engine room dip when the captain went into reverse. The move, he thought, had pierced the hull.

At 6:28, the captain ordered all engines stopped, ending what seemed like a thirteen-minute earthquake to the men in the engine room. They worked quickly to survey the damage and close every valve they could find. But water was still coming in. A major intake line had broken.

The chief engineer talked to Papadopoulos by telephone to tell him they were taking on water. The captain decided his only recourse was to try again. He ordered the vessel into reverse once more. Once more, the engine room crew staggered in the vibration as the single propeller tried in vain to pull the tanker from the shoal that had snared her. But she was stuck fast. At 6:55, nearly an hour after the ship hit bottom, Papadopoulos gave up. The engines were shut down. He would have to put out a Mayday. The *Argo Merchant* was hard aground.

Although water was pouring into the engine room, none of the ship's cargo of oil was pouring into the sea. Her thirty cargo tanks, laid out like an ice cube tray in ten rows of three across, were intact. But the water in the engine room at the stern was already putting an unusual stress on the rest of the ship. The seas piled up twelve- and fifteen-foot waves which were battering her portside. She started to list to starboard. She was helpless.

The 7,677,684 gallons of oil that were sloshing around in her tanks had been destined to take the first chill of winter out of hundreds of New England homes, schools, and businesses. Now, encased in a wounded tanker, the oil was trapped atop a shoal somewhere between the sandy beaches and lucrative shellfishing areas of Nantucket and Cape Cod and the rich offshore fishing grounds of a section of the Outer Continental Shelf known as Georges Bank. All that restrained the thick, black goo from pouring into the sea was an inch-thick skin of steel. And each surging wave slowly raised the bow, then let it drop with a thud against the hard, sandy shoal.

2

The Search

Lieutenant Commander Barry E. Chambers left his home in Elizabeth City, North Carolina, about seven o'clock on the morning of December 15. He was headed for a meeting in Norfolk, Virginia, to discuss a rash of recent oil spills in Chesapeake Bay. As commanding officer of the U. S. Coast Guard's Atlantic Strike Team, Chambers had played an important role in cleaning up the oil which had covered miles of shoreline on both sides of the bay with a thick gooey blanket.

He could speak at such meetings not only as an experienced supervisor of cleanup operations at an oil spill, but as an officer who had felt the squish of oil through his fingers, who had maneuvered thousands of feet of boom in efforts to capture spilled oil, and who had spent hundreds of hours scrubbing himself to clean black stains from his body. He had come to accept these conferences as part of his job, but he was much happier sitting in a pool of oil than sitting around a table talking about it.

At 8 A.M., he walked into the Fifth Coast Guard District headquarters in Norfolk. He had no sooner poured himself a cup of coffee when the district's pollution officer came up to him.

"Mr. Chambers, you better call your office. There's a tanker aground on Nantucket Shoals with millions of gallons of oil on it. Your team is already going."

"Already going," Chambers repeated half to himself. As he headed for a telephone, he wondered if the team had left without him. Though he is leader of the unit, the other officers and enlisted men have enough training and expertise to load up an airplane with special pollution control gear and report to a spill without him. At eight in the morning, the Strike Team would be arriving for work, Chambers thought, so nearly everyone would be there. He had no doubt preparations would go smoothly. He didn't want to get left behind.

At thirty-five, Chambers is a lean, sturdy man of medium height. He has sharp, penetrating eyes that squint almost shut when he smiles. And whether he is stern-faced while pondering his strategy at an oil spill or relaxing at a bar, he projects self-confidence. He wears black cowboy boots with his Coast Guard dress blues, and there is a cockiness to his gait especially when he is not in uniform that suggests just a trace of show-off.

When he came to the Strike Team in 1973 as its executive officer, the second in command, it was little more than a telephone in a cubbyhole of a bachelor officers' quarters. Now it is a full-fledged unit that occupies two buildings and maintains several million dollars worth of equipment, and Chambers has been a key to this transformation. If a lot oil spills or is about to spill anywhere along the East Coast or inland north and westerly to the Great Lakes, Chambers and the Strike Team get the call.

When he phoned his office, he learned that his men had just been notified of the tanker and were starting to load a C-130 cargo plane from the other side of the air station. They would leave at 10:15. Chambers gulped the rest of his coffee, went to find the district commandant, and excused himself from the meeting.

He immediately started charting his steps between Norfolk and his headquarters. He had had to park three blocks away from the building, in a tight space. It would take him a few extra seconds to free the car. The morning traffic would be coming into Norfolk, but he would be going out. The fifty-seven-mile highway route back would be free of tie-ups. And he remembered that he needed toothpaste.

The day before, he and some other members of his twenty-man team had worked on an oil spill in the Potomac River at Quantico, Virginia. He had returned home near midnight and before going to bed had methodically done a laundry and repacked his "response gear," including a uniform and some work coveralls he wears at spills. He had failed to do that once and wound up reporting to a diving job in a suit coat and tie and then swimming in his underwear. This time he had rummaged through his whole house, searching in vain for a tube of toothpaste. He knew he would have to get some before he got back to the base.

Leaving the building, Chambers felt the excitement of an oil spill alert pumping through his veins. The best he could do on the way to Elizabeth City was eighty-five miles an hour.

He had not yet had time to consider the stranded tanker itself. But plenty of other Coast Guardsmen had already been working on that problem for more than an hour.

When the distress call came into the Brant Point Coast Guard station on Nantucket Island, twenty-year-old Ken Larson was alone in the radio room, reading a book on navigation. It was just after 7 A.M. Out the window behind him, the sunrise was beginning to brighten the sky over the waters of Nantucket Harbor. As part of the routine of the man on the 4 A.M. to 8 A.M. watch, Larson had just roused eight other Coast Guardsmen who were sleeping on the second floor. He had stepped quietly down the stairs and eased back into his chair in front of a silver-gray microphone that was as big as his fist. It was suspended on a retractable steel arm over a bank of radios that had been silent for the first three hours of his watch. The only sounds now were the turning of the pages and an occasional creak in his chair.

"Mayday!" The voice cracked over the radio, shattering the stillness in the station. Larson snapped forward in his chair.

"Mayday!" the voice repeated. "Mayday! Mayday!"

Larson hit the foot pedal under his desk to activate his microphone and leaned over to speak. "Vessel calling Mayday, please send your position."

Other Coast Guard stations on Cape Cod and Long Island, New York also had picked up the distress signal and were trying to respond as well. In the cacophony of radio traffic, Larson picked out a voice identifying the source of the distress as a vessel named the *Argo Merchant*. The voice estimated its position as nine and one half miles south of Nantucket.

"Break! Break! All stations!" Larson barked into the microphone. "This is Brant Point Coast Guard. This vessel is in my area." The radio was silent for a moment. "Break! Break!" he continued. "*Argo Merchant*, this is Brant Point Coast Guard. Send your position in latitude and longitude."

While he awaited the vessel's response, he hollered to Ken Porter, the officer on duty, that he had a Mayday going. Porter, who was getting a cup of coffee at the far end of the next room, rushed to the radio room. When he learned the vessel was nine miles off the island, he didn't wait for more information. He ordered the station's forty-four-foot rescue boat underway immediately. Four men, barely out of bed, grabbed wetsuits and bolted out the door. They could dress on the way. Then Porter called the staff duty officer at Coast Guard Group Woods Hole, Massachusetts, the man in charge of search and rescue missions for a chain of nine small-boat Coast Guard stations along the southeastern New England coast.

Larson and Porter both figured they had a fishing vessel out there. Two had grounded on Nantucket Shoals during the past two months. Besides, no one but a fisherman braves the frigid, often stormy waters off New England during the winter.

The coordinates came back garbled but seemed to establish a position nine and one half miles off the island. Larson worked for more information.

"*Argo Merchant*, what is the nature of your distress?" he inquired.

"The *Argo Merchant* is aground and taking on water. We are in danger of capsizing."

"How many people do you have on board?"

"Thirty-eight."

"Thirty-eight!" Larson was startled. Most sixty- to eighty-foot fishing vessels have a crew of five at the most. "My God, what is this?" he wondered aloud. Though it seemed unlikely that any charter fishing boat party would be off Nantucket in mid-December, he could think of no other possibility.

"Okay, *Argo Merchant,* how big is the boat?" he continued.

"It's a very big boat."

"How many feet long is it?"

"It's a very big boat."

Larson insisted on the precise measurements. It was several minutes before the radio operator unrolled the vessel's drawings and found the information.

"Coast Guard, the *Argo Merchant* is six hundred forty-one feet."

Larson and Porter looked at each other. Their jaws dropped. Any notion about fishing boats and fishing parties vanished. They had a major problem on their hands. The answer to the next question put it clearly in focus: the six-hundred-forty-one-foot ship was full of oil.

Neither man had ever handled a distress call from an oil tanker before. But they treated it as they would any search and rescue: they sought more information. Most important, they wanted to confirm the vessel's position to dispel doubts raised by the fuzzy transmission minutes before. Larson asked once again for the coordinates. The reply came back more than thirty miles from the original position.

"My God," Larson said. "He doesn't know where he is." He asked for the position once again. It came back different from the other two.

"Okay, *Argo Merchant,* do you have Nantucket on your radar?"

"No. Nantucket is not on the radar. Nothing is on the radar."

Larson and Porter were convinced the tanker was more than nine miles away. A ship that large would have to be equipped with a radar that could easily pick up land within that distance. So they were stymied. They had a boat underway, though the forty-four-footer would look like a dinghy beside an oil tanker. They also knew Coast Guard helicopters from the Coast Guard Air Station on Cape Cod would be needed to lift the imperiled crewmen from the tanker's deck. But both the boat and the aircraft needed directions. The *Argo Merchant* was listing off a series of positions that covered several hundred square miles of ocean. If the vessel was in imminent danger, an accurate position was critical to the lives of the crew.

Exasperated, Larson asked the ship to send a position in loran bearings instead of latitude and longitude. The *Argo Merchant*'s radioman didn't know what Larson was talking about. Loran, an acronym for Long Range Navigation, is an electronic device that enables vessels to pinpoint their positions without the aid of the stars or other navigation instruments. On oceangoing ships of more than one hundred feet, a loran set, though not required, is as standard a piece of equipment as a radio on a car. Larson couldn't believe the *Argo Merchant* didn't understand. He spelled it out in the international mariner's alphabet: "Lima, Oscar, Romeo, Alpha, November." At first, that didn't help. Finally, the radioman said the ship didn't have loran.

There was little more Brant Point could do. But the radio traffic plus an almost constant telephone relay to Woods Hole had already set in motion a large-scale air and sea search and rescue operation that was taking on the urgency of war.

At 7:10 A.M., Lieutenant Christopher Burns was flying a Coast Guard C-130 cargo plane over the coast of New Jersey when he and his copilot happened to pick up snatches of a radio conversation about a vessel in distress somewhere near Nantucket. They were enroute to Groton, Connecticut, where they were to pick up the U. S. Coast Guard Academy Band and transport it to Washington, D. C. At first, all Burns and his copilot heard was the Coast Guard side of the conversation. But it was enough to indicate it was a Mayday. Under Coast Guard procedures, any emergency call takes precedence over any other mission. Routine flights frequently are diverted to assist on search and rescue calls.

By 7:20, Burns was close enough to hear both sides of the

conversation. He radioed Brant Point and confirmed that no other Coast Guard units were at the scene. Without waiting for orders, Burns told his copilot to turn easterly, toward Nantucket. And he began talking to the *Argo Merchant.* The tanker's radio operator gave a position southwest of Nantucket.

Meanwhile, when the staff duty officer at Group Woods Hole had enough information on the case from Brant Point, he called the Rescue Coordination Center for the First Coast Guard District in Boston on a special search and rescue hotline that links several Coast Guard installations at once.

At the Coast Guard Air Station just north of Falmouth on Cape Cod, Lieutenant Commander Bill Fisher, a helicopter pilot, had just walked into the station's search and rescue center, the headquarters of his unit's twenty-four-hour-a-day, thirty-minute standby alert rescue operation. He heard a radio speaker crackle with the Woods Hole-to-Boston message.

"Brant Point is working a motor vessel aground off Nantucket," the voice announced. "The vessel is reported in danger of capsizing."

Fisher raised his eyebrows. A fishing boat might tip over, he reasoned, but a "motor vessel" is a large ship, generally not prone to capsizing. He figured the message must have been garbled. The man at the Boston end of the network had similar doubts.

"How big is the vessel?" Fisher heard him ask.

"Six hundred forty-one feet."

"And she's about to capsize? You've got to be kidding."

"Well, that's the report, sir. And she's got thirty-eight POB's."

Thirty-eight passengers. Capsizing or not, lifting that many people off the boat would be a major rescue. Fisher ran out of the building to the adjacent runway and prepared to launch a helicopter.

At 7:29, Fisher lifted off in the first helicopter. Within ten minutes two more were airborne. All headed for the most recent position they had, thirty-nine miles southwest of Nantucket. It was the same area Burns was searching in his C-130.

Burns was trying to locate the ship with special homing

gear that locks onto an FM radio transmission and swings a dial on his instrument panel to the heading of the broadcast source. But the signal in that area from the *Argo Merchant* was too weak. The vessel was not around there. Burns asked the tanker for more information.

Papadopoulos had given the radio operator four coordinates to send out in the Mayday call. They traced out a patch of ocean thirty-six by thirty-eight miles—nearly 1,400 square miles—to the south and east of Nantucket. The captain had guessed he was somewhere in there. But none of the Coast Guard people monitoring the distress had understood that the tanker was trying to send a general area instead of a specific location until Burns picked it up in the airplane. He plotted the coordinates on a map he had propped up on his knees and headed his plane to the east.

Gradually, the signal grew stronger. By 7:35, the homing device began to respond. So that the transmission would not be interrupted, Burns asked the radio operator for a long count— to count slowly to ten and back. The needle on the dial swung to an easterly heading. The copilot turned the plane until its course and the homing dial heading were the same. The C-130 flew toward its target.

In the waters forty miles southeast of the Nantucket Lightship, the Coast Guard Cutter *Sherman* was steaming northward that morning. She and her crew of 172 were on a routine offshore fisheries patrol, enforcing regulations that govern the American and foreign fishermen who harvest the rich fishing grounds of Georges Bank. The *Sherman* was on a northwesterly course, headed for one more detail—to check the deep sea scallopers off Pollock Rip, between Nantucket Island and Chatham on Cape Cod—before returning to Boston, her home port, in time for Christmas.

In the *Sherman*'s radio room shortly after seven, Steve Davis picked up an SOS signal, followed by a message in Morse code, announcing a tanker in distress. The code was uneven and erratic.

"The guy must be nervous," Davis thought.

The code listed the coordinates of the tanker's position. Davis took them down and showed them to the officer on watch. Both knew immediately the position was incorrect. On a chart, it showed up in the Gulf of Maine. On the second try, the coordinates pinpointed the vessel in the state of Idaho.

When they were finally able to plot a reasonable position, they notified Capt. Michael Veillette, the skipper of the *Sherman*. He looked at the chart and figured the SOS would be someone else's problem. The tanker was two miles off Nantucket Island. The *Sherman* is 378 feet long and has a navigational draft of twenty-two feet, far too deep to sail safely in water so close to the island. But it occurred to Veillette, as he looked at the rash of light blue patches on the chart that marked the shallow water around the island, that if the *Sherman* couldn't rush to assist, how did an oil tanker manage to get in there in the first place?

Within minutes, the *Sherman* received radio orders from the First Coast Guard District in Boston to head toward the tanker anyway. The cutter had a deck large enough to land the Coast Guard's biggest helicopters. It was equipped with pumps that could be airlifted to the stricken vessel. And she was the largest vessel in the area. Boston wanted her to be the "onscene commander," the unit in charge of the rescue operations.

Veillette ordered the *Sherman*'s second turbine turned on. The cutter surged forward, toward its top speed, twenty-two knots.

Meanwhile, about sixty miles southwest of Nantucket, the Coast Guard Cutter *Vigilant,* one day out of New Bedford, Massachusetts, was also on an offshore fisheries assignment. Her officers and crew learned of the stranded tanker at 7:20, through a radio call from Boston that the watch officer picked up on the bridge.

"*Vigilant,* we have a report of a six-hundred-foot tanker aground and taking on water eight and one half miles south of Nantucket. Please proceed and assist. We'll get you an exact position later."

The watch officer immediately picked up the bridge telephone to inform Commander Ian Cruickshank, the vessel's skipper, who was having breakfast below.

"Okay," Cruickshank replied. "You'd better head over there, BTW."

BTW is a bit of Coast Guard vernacular meaning "balls to the wall." For the *Vigilant,* that is about fifteen knots, a little slower in heavy seas. A southwest wind had blown nearly forty knots all night. The seas were running twelve feet. Whatever it was that had run aground, Cruickshank thought, it was in some sloppy weather.

Flying in the C-130, Burns was closing in on the tanker. The homing gear was locked on to a strong signal that was kept loud and clear through repeated long counts from the ship.

At 7:50, Burns spotted the vessel below him.

"There it is!" he said into the microphone. "We've got him."

From five thousand feet, the vessel was a mere speck on the vast ocean. The only indication that it was in trouble was something Burns couldn't see: no wake was fanning out from its stern. The tanker was dead in the water.

As Burns circled the vessel, descending to get a closer look, he estimated its position using the navigation instruments aboard his plane. It was thirty miles southeast of Nantucket, he figured. He radioed the information to Lieutenant Commander Fisher, who was aloft in the helicopter.

Then Burns considered dropping an emergency pump to the deck of the tanker by parachute. But he knew the tanker couldn't move. The pump would have to land right on the deck, a tricky feat in calm weather, not to mention thirty- to forty-knot winds. The helicopters had pumps as well, and they could lower them to the deck on a cable, he reasoned. He decided to wait.

When Burns's first search southwest of Nantucket had turned up nothing, Fisher had wondered whether the Mayday was a hoax. The image of an oil tanker capsizing had seemed even more ridiculous than when he had heard the first report back at the air station. But the latest message from the C-130 dashed those thoughts. He turned his helicopter eastward, picked up the tanker's radio transmissions on his own homing gear, and headed for the ship. He was there in nine minutes, at 7:59.

When he looked at the tanker, Fisher knew immediately why someone on board had feared it would capsize. It was sideways to the seas, rocking back and forth. Huge waves broke over the decks, burying the port side in white foam. As he swooped down into a hover off the stern seventy feet above the water, he saw crewmen on the deck, their chests bulging with life jackets, their necks wrenched upward toward his helicopter and its promise of safety. They had been there for nearly an hour, shivering in fear and scanning the horizon in a vain search for help. They looked eager to get off.

But Fisher had another idea. Now that a helicopter and an airplane were overhead, he felt the psychological condition on the tanker had changed. He told the captain he could lower a pump to the vessel. Several more were on the way. The crew, he suggested, could try to get some water out of the engine room.

He guessed right. The captain agreed to take the pumps and try to control the flooding. And he told Fisher to forget about an evacuation. The crew would stay aboard.

Two crewmen in the back of Fisher's helicopter rolled an orange canister about half the size of an oil drum to the cargo door on the starboard side of the aircraft. They attached it to a cable and lowered the package to the deck. Inside the canister were a pump, a gasoline motor, rigging and hose—everything required to get the pump in operation.

But Fisher also knew that with an oil tanker aground, the pump would be as useful as a garden hose at a house fire. Known as a P-60, its capacity is sixty gallons a minute, 3,600 gallons an hour. It is designed for vessels up to eighty feet. Even a half dozen aboard might only be sufficient to keep some control of the water in the engine room until larger pumps could be deployed. In addition, the P-60's can't pump oil. Fisher thought from the start that oil might have to be moved before the operation proceeded very far.

As the pump was lowered to the deck, Fisher perused the

tanker, searching for places to deliver packages substantially larger than a small orange barrel. He thought there was room.

"I think we ought to get the Adapts in here," he radioed to Burns. "We could bring them into Nantucket."

Adapts are high volume pumps, capable of moving both water and oil. They are maintained and operated by Chambers' Strike Team in Elizabeth City and are designed specifically for combatting oil spills.

From his cockpit, Burns made a telephone call directly to the Coast Guard Commander of the Atlantic Area in New York and advised him that the Atlantic Strike Team and the pumps were needed on Nantucket.

By 8:10 A.M., one hour and five minutes after the Mayday and twenty minutes after the *Argo Merchant* was spotted, two more helicopters had joined Fisher and Burns circling overhead. Two Coast Guard cutters, one to the west and one to the southeast, were steaming toward her. And more than five hundred miles away, Chambers was speeding out of Norfolk toward Elizabeth City, where the Atlantic Strike Team was mobilizing for a battle to save her oil.

3

The Response

The *Argo Merchant* had run aground on the southern end of Fishing Rip, one of the numerous underwater mounds that extend southeast from Nantucket Island and make up what are known as the Nantucket Shoals. Fishing Rip, which runs twenty-six miles north-to-south and is six and one half miles at its widest point, is one of the outermost of the concentric, crescent-shaped shoals, which reach twenty-three to forty miles into the North Atlantic. Some of the ridges that are part of this geological formation, the Ice Age-old tracks of a receding glacier, lie just three or four feet beneath the surface, while depths at others range from fifteen to sixty feet. The outer shoals are separated by narrow, unmarked channels where the bottom drops off like a cliff and where currents run like river rapids, constantly carving, reshaping, and wearing away the sand.

On calm days, the shallowest ridges of Fishing Rip are visible from the air, spreading out under the water like giant tree limbs stripped of their foliage. But they are actually easier to mark in breezy weather, when the swells, rushing in from deep water on wind and tide, run abruptly into the shallow obstructions and pile up on each other until they fall over like breakers on a beach. The breaking line of white water on an otherwise deep blue surface traces the contour of the shoals.

On this morning, however, it was too rough even for that.

The twenty-five to forty-knot winds had kicked up so much white water that it was impossible to distinguish the waves breaking on the shallows from the other white caps. The eight- to ten-foot swells out of the southwest were building up to twelve to fifteen feet and higher on the shoals before smashing into the tanker.

The ship was located twenty-seven miles southeast of Nantucket Island and thirty-one miles almost due north of the Nantucket Lightship, which the captain and his officers had so urgently hoped to see, and which, until the moment of the grounding, Papadopoulos had insisted was still to the north of his vessel. Hundreds of tiny numbers dot the nautical chart of the area, giving the navigator a detailed depth map of the bottom. The number that falls nearest the spot where the tanker was perched reads three and one-half. Measured in fathoms, that is twenty-one feet.

Other charts, kept by historians, framed by sea-lovers, and sold to tourists on Nantucket, don't show the tiny depth readings of the navigator's chart. But they are even better testimony to the depth of the water there. These charts mark the graves of more than seven hundred ships that have run afoul of the shoals during the past three centuries and have never come off.

Captain Walter Folger keeps one of those charts on a wall in the living room of his Nantucket home, and on this morning, as news of the stranded tanker reached him in his Boston office, he had it pictured clearly in his mind. As chief of the marine safety division of the First Coast Guard District, Folger headed what is known as the regional response team, a special group of federal officials that meets in the event of major oil spills to provide advice and equipment to the people working at the scene of the spill. As Folger set in motion a previously prepared plan to get those officials together, the chart and the story it told began to haunt him. He had never known a ship to free itself once it was stuck on the shoals. An oil tanker out there, he thought, would be catastrophic. He paused for a moment, imagined tons of black oil coating his native Nantucket's beaches, and hoped for a miracle.

Folger had met briefly with his staff at about 8:15 A.M. Then, as they headed for telephones to alert members of the response team, he got on the telephone himself, to call Captain Lynn Hein, captain of the port of Boston. In that position, Hein serves as the on-scene coordinator, the man in charge, for any U. S. Government action taken within his area to prevent, contain, or clean up an oil spill. If the *Argo Merchant* grounding turned into a pollution incident, Hein would run the show. Everyone, including the Strike Team and other federal agencies involved, would work for him. He would report to and get advice from the regional response team.

The Coast Guard's mission at this point, however, was "search and rescue." The tanker had just been located. Thirty-eight crewmen were in danger. The oil was a secondary consideration. Nevertheless, both men anticipated that the key element in the incident would be the vessel's cargo. Folger urged Hein to get a handle on the situation immediately.

"Why don't you get an overflight, find out if the ship is in danger of breaking up or what the problem is," Folger said. "Just find out what is going on."

When they hung up, Folger headed for a meeting with Rear Admiral James P. Stewart, commander of the First District, to urge that the Coast Guard invoke the Intervention Convention and take control of the tanker. The international agreement, one of several drawn up by a United Nations agency known as IMCO, the Intergovernmental Maritime Consultative Organization, would permit the United States to take over the vessel if it presented a "grave and imminent danger" to the coastline. Without the authority of the convention, any effort to salvage the ship or tow it off the shoal could be regarded as illegal interference with a ship on the high seas or even an act of aggression since the tanker was in international waters. The Coast Guard Commandant in Washington would have to decide whether to take such action. Folger thought it should be considered as soon as possible.

Meanwhile, Hein walked into what was to have been a routine meeting of his staff. Like Folger, Hein was also familiar with the shoals and knew from first-hand experience that the tanker was in a precarious spot. As skipper of a 205-foot Coast

Guard cutter a few years back, Hein had patrolled the area off the shoals and occasionally had maneuvered in the water around Fishing Rip to rescue fishing boats that were taking on water or had suffered engine failure. He remembered the vicious currents that had made the shallow water even more troublesome. Whenever he was near the shoals, he hardly took his eye off his depth finder, and he took a fix with his loran set every five minutes. As he considered the grounded tanker on his way to the meeting room, he envisioned globs of oil washing up all over Cape Cod.

"Gentlemen, we've got a tanker aground on Nantucket Shoals," he said. "She's full of oil. I don't know what shape she's in. I don't think she's leaking oil. I don't know much more than that. But we've got to gear up for a major response."

A unit of three officers and ten enlisted men, one component of Hein's command, work under him just to respond to spills in the Boston Zone. By Coast Guard boundaries, that stretches from Hampton Beach, New Hampshire, around Cape Cod and Nantucket to Buzzards Bay. The men, one of thirty-five field teams around the country, check out about three oil spill reports a week and usually have to clean one up every two weeks. Some of these spills are little more than a couple of gallons that were washed off road surfaces in a rain shower. Others are larger, spilling from tankers during discharge operations at a terminal. Most are mystery spills, where the source is never determined though the suspect culprits are tankers which illegally pump overboard the oily water that was used to clean their tanks. But since he had taken command of the office August 1, 1976, Hein had never had an incident serious enough to call the Strike Team. And he certainly had never had anything in his area close to the magnitude of a grounded oil tanker.

In November, Hein had spent two weeks at a special Coast Guard seminar on oil spills and what he learned was still fresh in his mind. If the *Argo Merchant*'s cargo had to be pumped out of her to get her off the shoals, he knew he would need a large ocean-going barge to hold the oil. He instructed his staff to start looking for one that was in the area and empty. He also suggested they prepare to move their command post for the incident from the office at 447 Commercial Street, Boston, to the air station on Cape Cod. He remembered that one of his men, Chief Boatswain's Mate Terry Cramer, was on his way to Nantucket that morning to visit the Brant Point station. He asked one of his men to make arrangements with the air station and with Nantucket Airport so Cramer could be airlifted out to the tanker and put aboard.

Meanwhile, Commander Dominick Calicchio, the executive officer in Hein's office, called the Amership Agency in New York, the managing agent for the *Argo Merchant*. He asked Captain Nicholas T. K. ("Captain Nick") Skarvelis what arrangements the tanker's owners were making to salvage the ship and prevent a pollution incident.

"Nantucket Shoals is a treacherous place," Calicchio told him. "The weather makes up pretty fast at this time of year. You had better get something going as soon as possible."

Skarvelis said he would get back to him.

Because the aircraft at the air station were all committed at the scene of the grounding, Hein was unable to fly from Boston to check out the stranded vessel. So about 9:45, he climbed into a Coast Guard car, switched on the set of blue emergency lights on top, and headed for the air station, eighty miles away, where he planned to catch the first available flight over the *Argo Merchant*.

Out over the stricken tanker, Lieutenant Burns in the C-130 was overseeing the rescue operations. He contacted the *Sherman* and the *Vigilant* by radio and learned that they both expected to be at the scene well before noon. He watched as the second helicopter arrived shortly after 8 A.M. and delivered its emergency pump to the deck of the tanker, and a third, smaller helicopter circled the ship while a Coast Guard photographer took pictures.

Burns also scanned the water around the tanker, looking for patches of oil that might be escaping from her tanks. He saw none.

At 8:30, Captain Papadopoulos informed Burns that two

pumps were in operation, moving water out of the engine room. But the captain wanted to go one step further. He asked permission to pump some of his cargo overboard with the vessel's cargo pumps to help lighten the ship. Burns denied the request. He told Papadopoulos to stand-by until senior Coast Guard officials in New York and Boston could consider it. Burns wanted them to know the situation before anyone deliberately created an oil spill.

Forty-five minutes later, the pilot got word back through New York that First District officials in Boston had turned down the captain's request. He relayed their response to the tanker.

With the intensity and confusion of the past two hours, Burns had thought little of Papadopoulos' request. His initial reaction to deny it seemed natural enough, and now it was corroborated by his senior officers. But as he was relieved of his duties over the tanker and headed toward Groton to pick up the band, he wondered why the captain had even bothered to ask. He must have known by then, Burns thought, that he was in international waters and that the Coast Guard had nothing to say about what he did with his cargo. And, Burns knew, international law permits the captain of a ship to take any measures he thinks necessary, no matter where he is, to save his ship and passengers.

It was Captain Folger who had denied the request, and he also knew his word was meaningless beyond the twelve mile contiguous coastal zone where the United States does have such authority. But as long as the captain had given him the opportunity to prevent any dumping of oil, Folger wasn't going to pass it up.

At the same time, Folger was taking action that would put the weight of international law behind his decision, if a bit retroactively. He met with the district commander to discuss the Intervention Convention. By 10 A.M., Rear Admiral Stewart was on the telephone to Washington, asking the Commandant to determine that the *Argo Merchant* posed a grave and imminent threat to the United States coast. That would allow Coast Guard and other federal units to take "reasonable" action to prevent or mitigate the threat of oil pollution. Stewart's call prompted discussions between senior Coast Guard officials, the State Department, and the Liberian government. Stewart also asked the Commandant to authorize use of the pollution fund, set up by Congress specifically to pay for federal efforts to prevent damage from oil spills.

Just after 8 A.M. Ian Cruickshank returned to the bridge of the *Vigilant* and began monitoring the radio conversations between the aircraft flying over the tanker. From the messages and directions from officials ashore, he knew only that the ship was aground and taking on water in the engine room. But he knew from his own experience, and from eavesdropping on the aircraft pilots, the Coast Guard had two chances to save the ship. If they were lucky, the tanker had just nosed its way up on a sandbar and was only lightly aground. Perhaps, Cruickshank thought, the captain had caught the vessel heading into trouble and had slowed down or even started backing up before it slid onto a shoal. It was a slim chance, Cruickshank knew, but if that were the situation, they might be able to tow the tanker off. If that were impossible, then the flooding in the engine room would have to be stopped and most of the water in there would have to be pumped out to give the vessel enough buoyancy to float off the shoal on its own or with help from a tow. That would require a damage control party. By midmorning, Cruickshank and the helicopter pilots were talking about getting a group of damage control specialists on the *Vigilant* prepared to go aboard the tanker.

In the Navy and the Coast Guard, damage control is a science and specialty all its own, in which men are trained to keep wounded vessels afloat. During World War II, while most of a Navy vessel's crew manned battle stations during a fight at sea, one special unit stood by to patch up holes, fight fires, pump out flooding compartments, and repair other damage caused by enemy fire. The rest of the crew, theoretically, could keep fighting.

In the Coast Guard, damage control specialists have evolved into key personnel in search and rescue missions, especially in aiding the crew of grounded or leaking boats. The

Vigilant, the *Sherman,* and other vessels keep an assortment of pumps, hoses, patching tools, fire extinguishers, and other gear aboard for aiding stricken vessels. Most of the cutters, however, carry equipment suited to pumping out the bilge of a hundred-foot fishing boat. The *Vigilant* doesn't routinely rush to the aid of a six-hundred-foot tanker.

As his ship sliced through choppy seas toward the *Argo Merchant,* Cruickshank ordered four of his damage control men to prepare themselves and their gear for hoisting to one of the helicopters and a ride to the tanker.

Southeast of the wounded ship, the crew of the *Sherman* were doing the same. And they also hauled out a double-braided nylon towing hawser that was twelve inches around. It was laid out on the fantail, ready to be used if the tanker could be towed.

Barry Chambers sped into Elizabeth City about 9:30 and went straight to his house, four miles from the air station. He called his office again. The loading of more than 10,000 pounds of equipment onto the aircraft was proceeding on schedule. He took off the dress uniform he had worn to the Norfolk meeting and packed it with his other clothes. He pulled on a pair of coveralls, scrawled out a note for his wife, Cindy, an airline flight attendant who was out of town that day, jumped back in his car, and headed for the station.

On the way, he stopped at a food store and picked up ten packages of cigars, some pipe tobacco, and a couple pocketfuls of candy bars and cans of Vienna sausage. Conditions aboard the tanker, he figured, might make meal arrangements difficult. He also bought a tube of toothpaste.

By the time he pulled into his headquarters about one third of a mile inside the air station gate, it was 10:15. The plane was loaded. The wives of his men, who had been alerted by urgent telephone calls from their husbands two hours earlier, were there, having brought response gear in from home while the men loaded the airplane.

The air at the station was taut with excitement. When the men are not on the job at a spill, they spend hours at their headquarters repairing and rebuilding their equipment and, like athletes, training constantly for the next spill. When the call came into the station that morning, it was game time. Stomachs tightened up, pulses pounded, and the men moved quickly to their assigned tasks, testing themselves against the challenge of getting their equipment ready to go within two hours. On some smaller spills, when just a few of the men are going, they bicker goodnaturedly among themselves, battling for the one of the seats on the plane. Everyone wants to go. None of that was necessary this time. A tanker aground off Nantucket was a big deal. Although only half the team was getting aboard the first flight, it was likely there would be enough excitement on this job to go around. Those staying behind would load more airplanes before flying up later. Others would drive equipment to Cape Cod. And someone had to stay around to keep the shop open.

There was just one problem with the call that morning: it came ten days before Christmas. Some of the men and their wives by now had permitted themselves to think they would make it till Christmas without a big spill. Some had even dared to plan Christmas parties. Now those hopes and plans were dashed. They should have known better anyway. Most of the veteran team members had spent Christmas of 1975 aboard a tanker in Puerto Rico.

With the equipment on the aircraft, Strike Team members kissed their wives, wished them Merry Christmas, and climbed aboard.

Stacked in the cargo hold of the C-130 were five metal pallets loaded with three Adapts pumps, the Strike Team's mainstay. Adapts (pronounced AY-dapts) is a military acronym for an almost unmanageable mouthful of technology: air-deliverable-anti-pollution-transfer-system. It is a 400-pound pump, eleven inches in diameter, shaped like a six-foot-long cannon. In one minute it can pump 1,800 gallons of oil. It is the fastest portable pump in the world. Without it, the Strike Team on many of its jobs would be as useful as a mechanic without a wrench.

Each Adapts unit is packed on two nine-foot by seven-foot pallets that are loaded onto an airplane with a forklift truck. The pump, hoses, rigging and coupling gear, extra fuel, and tools are carefully stacked on one pallet. The "prime mover," a 1,200-pound diesel engine which supplies the power to the pump, rests on the second pallet. The Strike Team keeps four complete units covered with canvas outside on an old runway adjacent to their headquarters, ready for loading. Three of them were now on the C-130.

The Strike Team also loaded a twenty-one-foot rubber inflatable boat known as a Zodiac, small salvage pumps, winter weather gear, walkie-talkie radios, emergency lighting systems, generators, tools, C-rations, and even toilet paper onto the airplane.

Once the men were inside the plane, however, the efficient loading procedures went for naught. The pilots checked out the loading pattern and weight and found they had more cargo than they had planned on. They had to take some fuel out of the plane, a painfully slow process, the team learned. They also had to check out the engines. Minutes turned to half hours. Chambers was getting edgy. December 15 is one of the shortest days of the year. To get a good start on the tanker, he wanted at least to see it in daylight.

It was nearly noon before the plane was ready. But the weather was poor, and another aircraft was just beginning an instrument landing approach to the air station. It took precedence over all other traffic. Takeoff was delayed again. At 12:32, two hours after the team was ready, the C-130 finally roared down the runway and soared into the air.

Within minutes, the pilot was notified by radio that a small plane had crashed near Richmond, Virginia. He diverted the flight to the scene. No matter what was happening to the oil in the tanker stuck on Nantucket Shoals, Chambers and his team might just as well have been the Coast Guard Band. The Coast Guard's first job is saving lives, not preventing oil spills.

After circling the crash scene for twenty minutes, the pilot was informed that other units had control of the incident. He was released. The C-130 and the Atlantic Strike Team veered back on course, finally heading for the Coast Guard Air Station, Cape Cod.

Terry Cramer of Captain Hein's office landed at Nantucket Airport about 9 A.M. on his way to Brant Point. His assignment was to instruct the station personnel in the initial handling of oil spills that might occur in inlets and harbors around the island. Under Coast Guard regulations, Hein's office is charged with responding to any reported spill within an hour. To accomplish that, he has to rely on outlying stations in some areas, particularly on Nantucket, to get to the spill before his specially trained oil pollution unit can get there.

But Cramer wasn't going to get to Brant Point that day. He had a message to call his office waiting for him as he stepped off the plane. His instructions were to take the next available helicopter out to the tanker.

About the same time, one of the small helicopters that had been over the tanker throughout the morning landed at Nantucket to refuel. Cramer had his ride. At 10:15 he was airborne. Twenty-five minutes later, as the pilot maneuvered into a hover about fifty feet above the stern of the ship, Cramer crouched into an iron rescue basket that was attached by cable to a winch. Crewmen eased the basket out the cargo door and Cramer was lowered to the deck.

The crew on the tanker seemed curious and fidgety when Cramer arrived, though he detected some relief in their eyes as he hit the deck. He was ushered immediately to the bridge in the pilot house amidships. As he walked along the iron-grating catwalk that ran along above a maze of cargo pipes between the deck house aft and the bridge he could hear the hissing of air leaking from the pipes as the ship rocked in the waves. The lines must be rusted, Cramer thought. It indicated to him that the tanker was not in good condition.

On the bridge, he talked briefly with First Mate Ypsilantis before establishing radio contact with the helicopters and Coast Guard cutters that were converging on the tanker. As he stood on the bridge, he could feel the bow heave up with the surge of the ocean swells, then sink back down and strike the ocean bottom with a thud.

At First District headquarters, Captain Folger and Rear Admiral Stewart were working out the details of their plans to invoke the Intervention Convention, anticipating approval from Washington. By noon a message was drafted for Captain Papadopoulos. It said:

1. It is understood that your vessel is aground and in danger of breaking up.
2. You have requested permission to discharge your cargo and that permission has been denied, repeat denied.
3. The Commandant of the Coast Guard is presently considering whether your situation constitutes a grave and imminent danger from oil pollution to the coastline and related interests of the United States. If the Commandant decides it is necessary, he may order any of the following actions.
   a. Action to remove or eliminate the pollution danger,
   b. Salvage operations, or
   c. Remove and if necessary destroy your ship and cargo which are the source of the danger.
4. Coast Guard units under my command from the First Coast Guard District will provide on-scene assistance in order to try to prevent the need for pumping your cargo into the sea. If it is determined that it is necessary to carry out any of the alternatives described above, you will be notified.
5. The prohibition I have ordered against your discharging cargo remains in effect.

The message was delivered via the *Sherman* early in the afternoon.

By then, however, the question of whether Papadopoulos would pump oil overboard had long been moot. Within two hours after the grounding, the flooding in the engine room had forced the engineer to shut down the ship's boilers and fill them with fresh water to prevent contamination by the inrushing salt water. The only power aboard was from an emergency generator which put out enough electricity to keep lights and other electrical equipment running, but not enough to operate the tanker's large cargo pumps. They were shut down as well.

But that presented the Coast Guard with another difficulty. Not only were the pumps shut off, but also the heating coils, which lined the vessel's thirty cargo tanks and kept the oil warm. Number Six oil is carried at about 120 degrees Fahrenheit to keep it fluid enough to pump easily. As it begins to cool, it thickens. At eighty degrees it has the consistency of peanut butter, thick enough to clog the best pumps.

Before Chambers, his team, and his pumps even got off the ground in Elizabeth City, the near-freezing temperatures and the forty-degree sea water had already started working against them: the oil was starting to cool.

4
The Damage Control Team

The first helicopter appeared over the *Vigilant's* stern about 10:45 A.M. to pick up members of the Coast Guard damage control party. The ten-foot swells made it too rough to land on the vessel comfortably, so the pilot hovered about twenty feet off the deck while two men and some of their equipment were hoisted aboard. Shortly after eleven, they headed off toward the *Argo Merchant,* nineteen miles away.

The pilot approached the stricken vessel from the port side, where the white hull seemed to ride slightly high in the water because the vessel was listing to starboard. He swung up around the bow, where the passengers could see the words "Argo Merchant" painted high on the hull in large black letters on a background of white, stained with rust. As the waves, deep blue and covered with white foam, slammed against it and passed beneath it, the bow nodded slowly and stiffly, as if trying to turn itself away from a beating it could not escape.

Looking down the deck, the passengers could see the huge steel-gray cargo pipes that ran lengthwise down the middle of the deck for about 150 feet before running beneath the forward deckhouse. While many tankers have just one house that rises above the massive deck, the *Argo Merchant* had two, one forward, where the bridge was located, and one aft, over the engine room, where the smokestack was located.

As the helicopter came around the bow, down the starboard side past the bridge, the men could see the waves crashing over from the port side, showering with a heavy spray the long, iron-grating catwalk that stretched about 250 feet between the two deck houses. The catwalk ran along the top of more cargo pipes, which continued from the bow section under the bridge and down the center of the deck almost to the engine room. The open deck space on either side of the cargo pipes between the bridge and the engine room would have been an ideal place to deliver the damage control team, but because of the list to starboard and the rough weather, much of it was awash with swirling sea water.

The pilot flew around the stern, over the blue and white striped smokestack that sat atop the afterhouse, and finally settled into a hover over the fantail, an open deck area at the stern of the ship. While the space was cramped and partially covered with a corrugated metal roof, it was the only safe place on the tanker to put passengers aboard.

The first two men from the damage control party were lowered one at a time in a rescue basket about twenty minutes before a similar team arrived from the *Sherman*. The helicopter returned to the *Vigilant* for two more men and more equipment. By 12:30, eight damage control personnel, four from each ship, and two large pumps were aboard the *Argo Merchant*.

While the helicopters were shuttling back and forth between the tanker and the cutters, the *Sherman* arrived on the scene and assumed overall command of the search and rescue mission at 11:15 A.M. "On-scene" for the 378-foot vessel meant within sight of the tanker about five miles to the west. With the helicopters and radio communication, Captain Veillette was in close enough touch with the people aboard the stricken tanker. He had no interest in venturing in too close and risking trouble with the shoals himself.

Shortly after noon, Captain Hein took off from the air station in an airplane to survey the situation. Hein knew it was Veillette's job to coordinate efforts to save the ship and the crew; he wasn't about to interfere. But he was gathering information about the status of the vessel in case the grounding evolved to a pollution incident, of which he would take charge. He was also filing an eyewitness report back to Rear Admiral Stewart, with whom he was talking on a direct telephone line.

His communication with the tanker, however, wasn't as direct. In fact, he couldn't talk to the men on the ship, and had to relay his questions and observations through the *Sherman*. The damage control party had just arrived and was busy rigging pumps. The men had not had time to get a good look at the vessel; thus the information they provided Hein was sketchy. His best evaluation was his own, made from more than twenty passes over the ship, and he didn't think it looked too bad.

From the air, the starboard list was obvious but didn't seem critical. For a grounded ship, it looked almost level. It did not appear to be bending from the stress of its position. And except for a slight sheen of oil from the water being pumped from the engine room by the small drop pumps, no oil was escaping.

"If we can stop the flooding, sir, and pump out the engine room, we might be able to get her off," Hein reported to the admiral. "If we can lighten her up, we could get a line on her and tow her off."

As he was leaving the scene to return to Boston, he asked Captain Veillette to have the damage control party take soundings of the oil tanks to see if any were leaking. And he decided he would like a first-hand report of conditions aboard the tanker from his own executive officer, Commander Calicchio, who had sailed as a master aboard commercial ships and who had several years' experience in marine safety and ship inspection.

Airborne in the C-130, Barry Chambers unrolled a navigational chart of the waters off Nantucket that one of his men had brought aboard and plotted the vessel's position with the coordinates the team had received from Boston. It was twenty-seven miles off Nantucket, and more than seventy-five miles from the air station on Cape Cod.

"Why aren't we going to Nantucket?" Chambers asked.

One member of the team had spent the morning on the telephone making advance arrangements for their arrival and had already checked out Nantucket while Chambers was enroute home from Norfolk. The island's small airport could accommodate a C-130, he told Chambers. But it lacked heavy equipment capable of unloading some of the heavy pallets of Strike Team equipment. That ruled out Nantucket and the air station was the only other reasonable choice.

Chambers, however, was not convinced. If necessary, his men could adjust to the limits of the island's equipment and break down the pallets to unload them piece by piece. It would be much easier, he thought, to airlift men and equipment to the tanker from twenty-seven miles rather than seventy-five. Captain Hein's officers had left the decision of where to locate up to the Strike Team.

It was a decision that involved not only the logistics of moving his team and his pumps but the problems of housing and feeding the team, getting rental cars, and finding hardware or supply stores. One time the team was sent to a spill on the Potomac River and the Coast Guard officials in charge suggested the men base their operations at a Naval facility in Indian Head, Maryland, on the eastern shore of the river. But Chambers had been to Indian Head and remembered it as pure redneck country. There wasn't even a McDonald's restaurant within thirty miles of the place. And on one night he had spent there on a diving job, he was bitten by bedbugs. On the other side of the river, however, Chambers knew a town was established around a Marine Corps base at Quantico, Virginia. It offered plenty of restaurants and hotel rooms, better boat access to the river, and a power and telephone hookup that was easily available. The team set up its command post at Quantico.

Even when local officials understand the needs of his unit, Chambers does not rely on them for much more than road directions. The team likes its independence and prefers to go into an area completely self-sufficient. Besides, Chambers reasons, on a major spill, the local people are preoccupied with their own problems and don't need to be burdened with chores like finding hotel space for a dozen extra men.

On first glance, the choice between Nantucket and the air station seemed to favor the island, because the tanker was so far offshore. But Nantucket was more than thirty miles over water from the air station, where the officer in charge of the spill would probably locate his headquarters, and the distance might hamper communications between him and the team. Chambers thought he would evaluate the situation when they landed at Cape Cod before deciding where his men would set up shop.

The Strike Team leader had little experience in the waters off Nantucket, and while he knew the weather was not good that morning, he had no idea what the water conditions were around the tanker. He wondered whether Boston officials were lining up a barge, whether any pollution contractors had been hired, and what Coast Guard vessels might be in the area. Most of all, he wondered about time. He already had lost two crucial hours on the ground at Elizabeth City. The limited information he had on the tanker's condition was so sketchy and had passed through so many hands first that it was almost useless to him. How long would it be, he wondered, before he would get aboard the vessel and judge it for himself?

As he studied the Nantucket chart, he noted that although the ship was far enough away to be a problem to get to, it was still close enough for its cargo to foul the beaches of Nantucket and Cape Cod. That he knew for sure.

"Well, guys," he remarked as his men broke out sandwiches they had brought aboard during the delay on the ground. "If she comes apart and that oil comes ashore, we'll be there till spring."

Once aboard, the damage control team from the *Vigilant* and the *Sherman* went straight to the engine room, the ship's cavernous cellar, fifty feet deep and eighty feet across the breadth of the vessel. Standing on an iron-grating platform on the main deck level, the team could look up twenty feet above them through another platform to the ceiling. Looking down, they could see two large boilers, a tangle of pipes, and a series of platforms, runways, and ladders that resembled scaffolding. Cleared of those items, the space would easily accommodate a large, three-story house. And as they looked down, as if standing on the roof, they could see the water which had inundated the room. It was covered with an oily sheen, flushed from the bilges by the flooding sea water. One of the men took a sounding, dropping a weighted line into the water: thirteen feet, six inches. Long strings of fire hose stretched upward from the suction nozzles of the six small P-60 pumps the helicopters had put aboard.

Out on main deck, water was trickling out of the discharge hoses. The P-60s are the size of lawn mower engines and they quickly lose their effectiveness if they have to move water more than twenty feet. The discharge hoses ran about fifty feet, mostly straight up. They were getting little water out. But at this point, according to the damage control party's initial appraisal, little water was coming in. The level in the engine room was constant.

The men thought they might get more water over the side by rigging three of the small pumps in a series so they would have the suction of three pulling water from deep in the engine room instead of just one. Using fire hose from the tanker, they linked three of the pumps and ran the discharge hose out one of the portholes on the listing starboard side. When they finally turned them on, the three pumps in tandem sucked a little more water from the engine room than they had individually. But the difference was negligible. The men started rigging the two larger P-250 pumps they had brought aboard. They were rated four times as fast as the drop pumps, but even that speed would be next to useless in the volume of water that had already made its way to the engine room.

After the three small pumps were hooked together, some of the damage control people went with Ypsilantis to inspect other compartments in the stern of the ship to see what the extent of the flooding was. Smaller compartments such as the steering and gear sections were dry, but the pump room, where the ship's large cargo pumps are enclosed just forward of the engine room, was a different story. Ypsilantis opened a hatch and started to climb down to the pumps when the Coast Guardsmen were staggered by the fumes. They stopped him. If anyone was overcome by the fumes, the men had no breathing equipment to revive him. They had to settle for a quick inspection by flashlight. Ray Meyer, a machinist technician from the *Vigilant*, shined his light down the hatch. He saw liquid. Just how far down it was, he couldn't tell. Nor did he know whether it was water that had leaked through from the engine room or pure oil that had leaked in from another source. Whatever it was, it was more than the oil and water that accumulate in pump rooms during normal pumping operations. It wasn't supposed to be there.

Meanwhile outside on the main deck, Terry Cramer was taking a tour of the ship. Shortly after 12:30, as he looked over the starboard side into the water, he noticed some globs of oil, coagulated like huge drops of black blood, floating to the surface from beneath the ship as it rolled in the seas. The *Argo Merchant* was beginning to leak oil.

As the on-scene commander, Veillette had two responsibilities: to make sure the crew was safe and to try to save the vessel. Once the damage control team got oriented aboard, Veillette asked Lt. Paul Hagstrom, the engineer officer of the *Sherman* and the ranking member of the damage control party, to assess the ship's condition and see whether it could be towed off the shoal. He also wanted Hagstrom to sound out Captain Papadopoulos on getting the crew off the tanker.

Hagstrom and his men were not concerned about the listing of the ship. It had nothing to do with any shift of the cargo or with the tanker's stability, they thought. She was not about to capsize. The list was caused by her position on the shoal. The waves, pounding the ship broadside from the port were pushing and grinding the hull deeper into the ocean bottom. Pulling her off from that position, even if it were possible, Hagstrom determined, would rip the bottom right off the ship and most of her cargo would gush into the sea.

The flooding of the engine room had not increased in the hour he had been aboard, a key factor, he thought, that gave them time to work with. Though the pumps he had were not likely to gain on the flooding level, he hoped they could somehow maintain the status quo until the Strike Team arrived with the high-volume pumps. If those men were able to empty the engine room, the ship would rise substantially out of the water, significantly improving the prospects of towing her off. He suggested to Veillette that any towing attempts wait until the Strike Team got aboard.

Just how long they could wait, however, was a critical question. Hagstrom and his men could see that the catwalk between the two deck houses had started to arch. The engine room was not sitting on the ocean bottom, as was the vessel's midsection. It was hanging off the shoal over deeper water. And the flooding in the engine room was affecting the hull of the ship the way weight effects the end of the diving board: it was bending it down.

About 1:30 P.M., the *Vigilant* arrived at the scene, having been delayed by maneuvers she had to make while the helicopters hovered over her deck. With her shallow, ten-foot draft, she could move to within a few hundred feet of the tanker. Commander Cruickshank started a slow, careful tour around the ship taking soundings of the bottom. He wanted a precise profile of the shoals around the tanker so he would know where he could maneuver safely and where the deepest water was so he and Veillette could determine which way the ship might be towed.

Cruickshank was wary of the charts for the shoals, thinking that the constant wave and current action around them might shift the sand. The shallowest reading around the grounded tanker was about twenty feet. Seas were running at least ten feet and in the troughs that left very little water over the shallow spots. With a little bad luck, Cruickshank thought, even his vessel might get stuck. He stationed extra people in key areas on the cutter, especially the anchors, because in addition to the water he was concerned about the *Vigilant*'s chronic engine control problems: the steering pump had a habit of failing, making the rudders useless. It would be a bit embarrassing, he thought, with the helicopters and airplanes buzzing overhead and the incident growing in magnitude, to ram the hull of the *Argo Merchant* in his first few hours out there.

Because of the dangerous water, it was unlikely that the heavier and more powerful *Sherman* could venture in close enough to get a line on the tanker. And there were no sea-going tugs in the area. Any immediate tow job, therefore, was up to the *Vigilant.*

To Cruickshank, the odds were overwhelming: his 1,000 tons against the *Argo Merchant*'s 35,000 tons. The *Vigilant* might as well have been a rowboat. He didn't think he could move her if the tanker was simply standing dead in the water, not to mention stuck hard aground on a shoal.

The skipper of the *Vigilant* learned from his tour around the tanker that the only good water was to the northwest, and that to have any hope of getting her off, the swift currents and the wind both would have to be behind her pushing as the *Vigilant* pulled on the other side. And Cruickshank figured that the only effective tow connection would be the ship's anchor chain, since that is tied directly to the tanker's structure. But before anyone could tie the *Vigilant*'s hawser onto the chain, the anchor would have to be removed, a formidable task, which would require people to work on the outside of the surging bow, suspended on a rope over the sea. And even then, the *Vigilant*'s eight-inch hawser would probably pull out a mooring bit on the cutter before the tanker would budge, or it would break, snapping back like a huge elastic, with such force it would wipe out anything in its path. Even with the precaution of clearing the decks on both ships, someone could get killed.

"I think it's impossible," he radioed to Veillette. "This thing's no fishing boat. We need a deep draft tug, with big propellers if we've got any chance at all."

But, for the moment anyway, the towing question was academic. When Hagstrom asked Papadopoulos about putting a line on the tanker to tow it off, the captain resisted. That would be up to the owners, he said. That was a significant consideration for Veillette. The Coast Guard hadn't taken over the ship yet under the Intervention Convention. Without that, the vessel was still under the control of her master. Without his permission, a towing attempt would be risky indeed. For if Veillette had ordered a line put on the vessel and had towed it off, only to watch it break up, spill its cargo, and sink, Veillette would have been in trouble: he would have sunk the ship.

If he thought he had a good chance of succeeding, Veillette was ready to take the risk. But the reports from Cruickshank and Hagstrom on the scene were bleak. It would not be towed, he decided, while the engine room was flooded with sea water. He would wait for the Strike Team.

Papadopoulos was as reluctant to release his crew as he was to authorize a tow. And once again, Veillette was hindered by traditional freedoms on the high seas: the captain is in charge of his crew. But Veillette was determined to evacuate some of the crew, for their own safety as well as that of the Coast Guard people on the tanker. In the first six hours of the incident, rather than take anyone off, the Coast Guard had put nine more people aboard. Instead of thirty-eight, there were now forty-seven. And with Strike Team and salvage experts expected, the plans were to put even more people on her. The last thing Veillette wanted was a floating hotel with a hole in the bottom rocking on that shoal. He knew the weather could get much worse with little warning. He had no idea how the tanker might withstand a good storm, but Hagstrom's report that some of the rails and catwalks were bending under the strain of the flooded engine room was a sign that it was starting to weaken already. Most of the crew weren't doing anything anyway, and if he could get twenty or thirty of them off when weather conditions were good, it would be twenty or thirty he and the helicopter pilots wouldn't have to worry about if problems arose later. Veillette wanted them off.

At 2 P.M., Papadopoulos finally agreed to allow half the crew to leave the ship. As the word passed among the crewmen, many of them disappeared to their rooms to collect their personal belongings and prepare to go ashore. Within fifteen minutes they started to emerge from below deck, some of them loaded down with suitcases and duffel bags. And they had changed their grimy work clothes and casual jeans. Instead, they were wearing bright sport jackets and ties. The Coast Guard workers were surprised and amused. The previously meek and frightened crewmen were transformed into sharp-looking characters, ready for a night on the town.

But they were only going to the *Sherman* for this night. And they weren't going with all their luggage. When the first helicopter pilot hovered over the deck and saw what the crewmen were hauling around, he radioed word to the deck that baggage was out. It would add extra, unnecessary weight to the aircraft and might limit the number of passengers the pilot could take on one run. When the crewmen heard the orders, however, they refused to let go of their belongings. Finally the pilot agreed to permit each of them one small suitcase.

As the iron rescue basket was lowered to the deck for the first time, the crewmen pushed and shoved each other like school children in a lunch line, eager to get to the front. While their captain had been reluctant to let them go, the crew had become more anxious to leave with each lurch of the ship. By now, some of them were desperate. Two of the youngest crewmen kept jumping to the front of the line in defiance of Chief Mate Ypsilantis, who was struggling to maintain order. One of them finally dodged the first mate during one of the hoists of the basket and ran out to grab it as it was coming to the deck. The basket, however, picks up a static electrical charge as it twirls downward and must hit the deck to discharge. When the crewman grabbed it, he got a jolt that threw him flat to the deck. Ypsilantis had no trouble with him from then on. In fact, it took some encouragement to get him into the basket when his turn came.

Within two hours, twenty of the crew had been airlifted in two helicopters to the *Sherman*. The fear in their eyes as they scrambled out of the helicopters onto the *Sherman*'s deck was gradually replaced by relief; they were glad to be aboard a ship that wasn't going down. At the same time, however, they were concerned about the belongings they had left behind. They asked the *Sherman* crew when they could go back to their ship. And they wondered about their money. Papadopoulos hadn't paid them yet for the voyage.

At 1 P.M., as Captain Hein was flying over the tanker, one of his staff members in Boston contacted the Marine Towing & Transportation Company in Philadelphia, which operates barge-towing tugboats along the East Coast. The company had a 70,000-barrel barge—capable of holding almost 40 percent of the *Argo Merchant*'s cargo—unloading that day at the New England Petroleum Company pier in Sandwich, Massachusetts, on Cape Cod Bay near the mouth of the Cape Cod Canal. It would be empty within five hours and available for towing to the scene of the grounded tanker.

It was a lucky find. Most barges plying the East Coast deal in "clean" petroleum products, the lighter fuels such as home heating oil, naphtha, and gasoline. They aren't rigged with heating coils needed to keep Number Six oil warm enough to pump out. And putting a "dirty" fuel in them would prevent their use again in the clean trade without substantial, costly, and time-consuming cleaning. This barge, however, routinely carried Number Six oil, and it was close by. Hein's office reserved the barge, the NEPCO 140, and the tug *Marjorie B McAllister*, pending confirmation that it would be needed at the scene.

Meanwhile at 2:35 P.M., Captain Skarvelis of the Amership Agency notified Commander Calicchio that a New York firm had been hired to handle salvage operations at the scene and that a pollution company was arranging to place a boom around the vessel to contain leaking oil and to take whatever measures would be necessary to protect the shoreline.

Calicchio hung up and headed for nearby Logan Airport, where he was to meet Captain Hein and get on his plane for a trip to the air station himself and a helicopter flight to the tanker. As Hein touched down at Logan shortly after 3 P.M., the teletype in District headquarters was clacking with a message from the Commandant's office in Washington:

1. The imminent threat of damage to the [*Argo Merchant*] and her cargo determined to present a grave and imminent threat to the coastline and related interests of the United States.

2. Authority granted to implement provisions of Intervention Convention Act if actions by vessel owners/agents considered by district commander to be inadequate, not timely or inappropriate to abate threat of pollution. Advise com-

mandant by most rapid means of action contemplated if this authority exercised.
3. Pollution Fund activated for federal action under this authority. Initial funding limit [$500,000]. . . .

Hein stepped off the airplane mildly annoyed. He had left his clothes and sleeping bag at the air station before he flew over the ship, anticipating that he would return to Cape Cod and perhaps spend the night there. But it was decided during the flight that he would head directly to Boston instead. Thus he would have to report to district headquarters in his coveralls.

But any concern over that inconvenience was put out of his mind when he got to headquarters and learned that three key weapons in his oil pollution fighting arsenal were in place: with the approval of Admiral Stewart, he had authority to take control of the ship; he had a barge nearly ready to sail toward the tanker; and he had the money to hire it.

And a fourth weapon was back at the Cape Cod Air Station. The Atlantic Strike Team had just arrived.

About midafternoon on the *Argo Merchant,* Ray Meyer and a couple other members of the damage control party were having a cigarette break. They had finished rigging two of their P-250 pumps and were ready to use them if the smaller pumps gave out or if the engine room started flooding again. With the water level constant, however, they decided to let the smaller pumps do the work and keep the big ones handy as a back-up.

The men had not been able to take the tank soundings Captain Hein had requested. Most of the main deck where the tank covers were was still awash with the sea as waves continued to pound the tanker. In fact, waves were crashing over the port side, which was more than eight feet out of the water. And Hagstrom, who spent a lot of time running from the bridge to the stern and back along the catwalk even got wet there, another ten feet above the deck. There was a small shelter about halfway down the catwalk and Hagstrom would make a dash for it, timing the waves so they would hit while he was under cover, then dash for the bridge house. He still got hit a couple of times. If one of them had caught him off guard, he thought, he could have been swept overboard.

The men, clad in orange wet-weather suits, kept warm easily while they worked. But when they took a break, sweaty from their work and wet from the sea, the wind sent a chill through them. Meyer shivered as he stood near the stern of the ship, smoking and looking down at the main deck where the tank covers were. He noticed that the metal had started to buckle. There was just a ripple on the deck between the last two tanks, and the hull itself seemed to have bent slightly to starboard. It looked like a banana. He watched as the waves formed relentless walls of water that slapped the tanker on the port side, raised the forward section off the bottom as they slid under the vessel, and bounced it off the bottom as they passed. Even near the stern, Meyer could feel the mid and forward sections of the ship settle back onto the bottom after each surging wave.

Shortly before 4 P.M., something gave way. Flooding in the engine room increased dramatically, without warning. And it wasn't just seawater. The oily sheen that had been on the surface was now a thick head of pure oil. Cargo was coming into the engine room. Within minutes, the small drop pumps coughed as they ingested the thick goo, overheated as they strained to pump it, and finally seized up, ruined. The damage control team scrambled to get the P-250s working, dropping seventy-pound suction nozzles into the oil, which had piled up to more than a foot thick. They floated. The oil would clog those pumps as well, the men knew, and they poked at the nozzles with long broom handles, trying to push them through, far enough into the water so the oil wouldn't foul them.

They were standing on the catwalk that was one level down from the main deck, about thirty feet above the floor of the engine room. But as they looked down, they were only a few feet above the oil-covered water. And it was coming in so fast they could watch it rise.

5

The Blackout

The Atlantic Strike Team touched down at the Cape Cod Air Station at 2:35 P.M. Barry Chambers scrambled out of the aircraft and glanced at the sky. At best, he had two hours of daylight left. He wanted to get moving. He told men to keep the equipment on the plane until he determined whether they would stay there or head for Nantucket. Then he went into the operations building at the air station for a round of discussions with officials there and telephone calls to the *Sherman* and to Boston.

The answers to some of the questions that nagged him on the airplane came quickly. Captain Hein's office had a tug and barge on standby. Pollution contractors had been alerted. The federal pollution fund had been activated. Chambers was delighted. And amazed. He had worked on other spills and had had to contend with what he called the "ostrich concept." It took days before anyone would decide on a path to take and a few more to decide whether to follow it. There are a lot of people around, he felt, who live by the axiom that if you don't make a decision, you can't make a mistake. But such people weren't running things in the First District. Hell, he thought, the Coast Guard learned the ship was aground at seven and had notified the Strike Team just after eight. It was incredibly fast.

The air station had set up a tentative plan for getting the

Strike Team's pumps out to the tanker: the units would be loaded from the C-130 onto flatbed trucks and transported down the road about fifteen miles to the Coast Guard station at Woods Hole, where they would be put aboard the buoytender *Bittersweet*. The *Bittersweet* would then steam out to the *Argo Merchant* and lift the pumps aboard the tanker. The plan was far enough along that Chief Warrant Officer Peter Brunk, another Strike Team member, had been informed of it and was already in Woods Hole, briefing the *Bittersweet* captain and crew on the kind of gear they would be carrying. Brunk had been in Hershey, Pennsylvania, that morning on a lecture assignment when he got a call from Elizabeth City. Within a half hour, he was on an airplane to Boston, where he rented a car and drove to Woods Hole. He was on Cape Cod before the rest of the team.

At first, Chambers agreed with the plan, and had his men start unloading the equipment from the plane onto the trucks. But as he gathered more information from his telephone calls with the *Sherman* and with Boston officials, he changed his mind. A ten-foot sea was running out near the grounding, he learned, much too rough to transfer the Adapts pumps from one ship to another. And he didn't want to wait the nine hours it would take the *Bittersweet* to get to the tanker, even if the weather were good.

"I don't think the *Bittersweet* route is a good one," he told people at the air station. "Once she gets out there, if she can't put that gear on board, I'm going to be stuck. It's going to be locked on that buoytender and I'm not going to have the flexibility of trying anything else. Why not take it out by helicopter?"

Lt. Cmdr. Carl Pearce, the assistant flight operations officer, was not enthusiastic. The largest of the two kinds of helicopters the station uses, the H-3, is capable of delivering the Adapts pump. But the pilot flight manual recommends it only as a last resort. The weight ratio between fuel and cargo is a key factor in flying a helicopter and the unusually heavy weight of the Adapts would displace some reserve fuel, a critical consideration because the tanker was so far offshore. The load was too heavy to be hoisted in and out of the aircraft; it would have to be carried underneath, attached with cables that would swing it like a pendulum. It would be dark before the first helicopter could get out there, and pilots almost never "sling load" packages beneath the helicopters at night. Finally, none of the air station pilots had ever handled the Adapts before, even in ideal conditions.

"You know what it's like flying a helicopter at night over the ocean?" Pearce asked Chambers. "It's like flying in a coffin. Everything is pure black. Our guys haven't even handled this Adapts kit in the daytime. Are you sure you need to get it out there tonight?"

Chambers nodded. "Well, yeah. We want to get going on that tanker. The engine room is flooding and we've got to get out there and pump it out. I can't do much about it sitting here on the beach. And you guys are my only way out there."

Chambers has often confronted hesitancy or apprehension not so much with words as with his presence. He projects optimism, and it becomes a subtle force in his favor. While the Strike Team never really says so, when the men report to a spill they seem to say, "Well, we're ready to do *our* job . . ." and let the people whose help they need silently finish the sentence. It either inspires them to action, or it dares them into it. Chambers had never had anyone deliver the Adapts pump at night, and he had little understanding of the hazards of night flying. But he didn't consider for a moment that the pilots couldn't do it. That just wasn't an option. Besides, other pilots had delivered the pump before in the day time, to open barges, and had just flown them out and plopped them down. No sweat.

To solve the distance and fuel problem the men discussed moving the Adapts gear to Nantucket and flying it by helicopter from there, as Chambers had thought originally. But that turned out to be impractical. The air station's two H-3s had been ferrying crewmen and Coast Guardsmen between the *Argo Merchant* and the two Coast Guard cutters at the scene all day long. One of them returned at 3:25 P.M. The other would not be back until shortly after four. The Adapts kits would have to be reloaded to an airplane, flown to Nantucket and unloaded. The helicopters would have to be checked and refueled, flown to Nantucket and fueled again. Chambers was about to lose his daylight and he hadn't seen the tanker yet. The Nantucket stop would only mean more complications and more delay. It would have to be run from the air station. Pearce agreed to try it, but he was not optimistic it would work.

Most of the pilots had worked all day over the tanker, and Pearce wanted a fresh pilot for the job. He could do it himself, but as the operations officer, he wanted to stay on top of the whole picture. Lt. Tom Burnaw and Lt. Bill Shorter were on search and rescue duty that night, and Pearce wanted them and their helicopter free of other responsibilities in case the tanker suddenly became unstable or broke up and people still aboard had to get off fast. In fact, he told those two pilots to fly out to Nantucket after dinner, where they would be forty miles closer to the tanker. The Adapts assignment fell in the lap of Lt. Rick McLean; it was almost by default.

Like the other pilots, McLean had never carried a sling load at night, and he had no idea how the Adapts would ride under his aircraft. And although he was a fresh pilot, that was a significant disadvantage in one respect: he hadn't seen the tanker and was not familiar with the stern, where he had to deliver the unit.

"I got a lot of negative thoughts about it, I'll tell you right now," he said to Pearce and Chambers.

"Well, we'd like to give it a try," Pearce replied. "Just take it easy, take your time, and give it a shot."

"I'll definitely give it a try. But I'm not going to endanger the lives of my crew or my helicopter by pushing it."

"What we'd like to do is pump out that engine room," Chambers explained. "It's flooding pretty badly and I'd like to get a shot at it tonight. We might be able to refloat the thing and tow her off."

"It just can't wait till daylight, huh?"

"Well, we got what, seven and a half million gallons of oil sittin' out there. If that son of a bitch breaks up, we might have oil all over Cape Cod. If we can't get it out there tonight, yeah, we still might have a chance in the morning. Then again, we might go out there and find she's broken up. And we'll have oil all over the place."

It was not really a question of trying or not trying. If for no other reason, McLean had enough pride and enough "the Coast Guard is always ready" spirit to get him out on the runway and into the cockpit. And he didn't want to see a major oil spill off Nantucket any more than anyone else. But at this point it wasn't an oil spill. It was a threat of an oil spill. And no one's life was in danger. He didn't quite agree that the delivery of that pump was as urgent as Chambers made it seem.

"Like I said, I'll give it my best," McLean said, "But I haven't flown a sling load in quite a while. I've never flown one this far. And I've never flown one at night. If I have any trouble with that out there, I'm going to drop it, without hesitation."

That was his ultimate safeguard. He could jettison the package into the sea with a push of a button in the cockpit.

"Okay," the pilot said. "What are the weights?"

In calculating the fuel and weight requirements, McLean and his copilot considered that they would be carrying a 1,200-pound diesel prime mover for the Adapts pump; four passengers, including Chambers, two of his team members, and a member of Captain Hein's staff; and all their gear. With the load underneath, the fastest he could fly without shaking and vibrating out of control was fifty-five knots, much slower than the 120 to 130 knots at which the large helicopters usually fly. That would lengthen the time of the trip and increase his fuel consumption.

He needed enough fuel to get out there, hover over the tanker while the load and the passengers were put on the deck, and get back to the air station, cruising in at 120 knots. Though it had been windy most of the day at the scene, it was now nearly calm, and that was a disadvantage for McLean. If he had a fifteen-knot wind blowing, it would act as a natural lift for the helicopter. McLean would need less power, and therefore less fuel, to hover over the ship. But there was no such luck. He stretched the fuel factor to the maximum, leaving the least possible safe amount for a reserve. If he had any doubts about making it back to the air station, he could land at Nantucket.

Before he went out to the aircraft, he talked with two pilots, Lt. Cmdr. Bill Fisher, the first pilot on the scene, and Lt. Cmdr. Tom Preston, who was the second.

"You're going to deliver that out there tonight?" they both asked, their eyes wide with amazement.

"Well, they said they've got to have it tonight, so I'm going to give it a try."

"I can't think of a worse time or a worse place to try to experiment with that Adapts kit than to deliver it at night to that thing," Preston said.

They told McLean the fantail was much too cramped, with a flagstaff, lifeboat, davits, and a small shed, for safe delivery of the pump, especially from a sling load.

"There's no way you're going to be able to get that thing on there," Fisher said. "I've already seen it, in daylight, and I wouldn't do it."

It wasn't what he wanted to hear, but it confirmed his first doubts about the mission. As McLean headed out to the aircraft, he was even more convinced he would never get the Adapts on the tanker.

The belly of the helicopter had to be rigged with a special sling assembly that included a twenty-five-foot line with a hook on the end for picking up the load and both a mechanical and electrical connection which enables the pilot to release the package from the cockpit. When the helicopter was fitted with the sling, McLean tested it out, pushing a button for the electrical release. It didn't work. The crew worked with it briefly, without solving the problem. But they were not concerned. The release often doesn't work properly unless a load is attached. And the mechanical release, a foot plunger near the pilot's seat, worked fine.

Shortly after 5 P.M., McLean and the helicopter were ready. It was dark.

The Strike Team was ready as well. Two pallets, one with the prime mover and one with the Adapts pump itself and its accessories, were rigged with a special shackle and harness, ready for the sling hook from the helicopter. As they unloaded their equipment, they also had tested one of the Adapts pumps and found it wouldn't work. Some of the men took it apart to find the trouble, while others unpacked another pump. That one worked, and it was now ready for airlifting to the *Argo Merchant*.

About 5:15, the helicopter with McLean and his passengers lifted off the runway, its sling hook dangling beneath it like a long, thin claw. It settled into a hover over the prime mover. A man on the runway grabbed the hook and attached it to the shackle on the load, then ran to get out of the way.

McLean gunned the machine to near maximum power, trying to move forward as he went up. They moved forward slightly. The payload dragged down the runway. They couldn't get off the ground. McLean eased off on his power, then tried again. Again, he dragged the package without getting into the air. The helicopter was too heavy. He came down slightly, making sure the package was firmly on the ground, and depressed the electrical switch to release the load. It stuck at first, and then finally released the shackle. McLean set his helicopter back on the runway.

"I guess you guys will have to go in the other helicopter," he said to Chambers. "We'll try taking the load out on our own. It's just too heavy for all of us."

Chambers was irritated. It had taken a long time to rig the helicopter, he thought, and now they couldn't get off the ground. It was not an auspicious beginning. Disgruntled, he grabbed his gear and walked off with the other three men toward the other helicopter.

McLean tried again, and this time picked up the load with power to spare.

"We're underway with the Adapts kit" he radioed to the air station's search and rescue center. It was nearly 6 P.M.

Inside the air station, Tom Burnaw and Bill Shorter, who had expected to spend the night on Nantucket, learned they wouldn't be going there after all. Instead, they would follow McLean out to the tanker with the Strike Team.

At 6:20 P.M., about thirty-five minutes behind McLean and the prime mover, Chambers was airborne, on his way to the *Argo Merchant*.

Several members of the Regional Response Team were already at Coast Guard district headquarters in Boston when Captain Hein returned from his flight over the tanker. The team operated out of a large conference room of the First District's Rescue Coordination Center on the fourth floor of an office building next to Boston's North Station. Team members from such agencies as the Environmental Protection Agency, the Fish and Wildlife Service, the Department of Defense, and the Coast Guard had immediate access to the message traffic coming into the center by teletype from vessels on the scene, from Washington, and from Coast Guard installations up and down the East Coast.

It was more a gathering of officials than a structured meeting. Members served as resource people for Hein, providing information and lining up services and equipment he would need to prevent, contain, or clean up an oil spill. Hein still wasn't officially in charge of the operation, but he assumed it would be only a matter of time before he was, and he wanted to be ready for it.

When he learned that the Commandant in Washington had authorized money from the pollution fund, Hein checked quickly with Amership in New York to see what arrangements the owners had made. No one had showed up at the tanker yet or had even arrived on Cape Cod. Hein immediately hired the barge and tug his men had put on reserve earlier in the afternoon, negotiating a contract over the telephone. The cost: $400 an hour. Before the barge, which was 465 feet long, could come safely alongside the tanker, however, large fenders would be needed to keep the two vessels from bashing together in the ocean swells. Hein asked the Strike Team to fly up two 5,000-pound rubber fenders known as Yokohama fenders, which were stockpiled in Elizabeth City.

Hein also wondered about the prospects of burning the tanker's cargo before it could escape into the sea. He knew the chances were slim; in his days of inspecting barges, he had walked on Number Six oil. When it is cold, it is just like asphalt. And even when it is burned in industrial furnaces, it has to be heated to more than 160 degrees and sprayed as an oil mist from nozzles before it will ignite easily or burn.

Jack Conlon, the EPA representative, confirmed Hein's hunch that burning the oil would be futile, especially if the oil cooled. No burning or wicking agents had been developed that promised success in lighting it. He also ruled out use of oil dispersing agents that, in the instance of the *Torrey Canyon* oil disaster off the British and French coasts in 1967, had caused more environmental damage than the oil itself.

Conlon and Hein also discussed the use of gelling agents, which, if mixed with the cargo, could harden it within the tanks and prevent it from spilling.

But it would take a pound of gelling agent to harden a gallon of oil. Just getting seven and a half million pounds of gelling agent was in itself out of the question, not to mention getting it out to the tanker, into the cargo holds, and finding a huge egg-beater to mix it with the oil.

It was apparent to Hein from the outset that his best chance of preventing a major oil spill was to keep the cargo in the only container available, the *Argo Merchant* itself, until enough water was pumped from the engine room and enough oil was discharged from her tanks into a barge that she was light enough to be towed off the shoal.

And at 5:45 P.M. those efforts were underway. Just as Lieutenant McLean was lifting the prime mover off the runway at the Cape Cod Air Station, a McAllister tug, towing the 465-foot barge, was leaving Sandwich, Massachusetts. The barge was expected to be at the scene of the grounding by daybreak.

The sun set over the *Argo Merchant* just after 4:30, and with an overcast sky, the horizon darkened quickly. Lights on the bridge, the after deck house, and around the deck illuminated the ship, tracing its outline boldly against the blackening sky while its reflection flickered in the water. The stricken tanker, which had taken a vicious beating throughout the day from the breaking waves, was now enveloped by an aura of peacefulness and beauty. Enblazoned on the black water and against the black sky, she actually looked pretty to the crews aboard the *Vigilant* and the *Sherman*.

The scene belied the frantic efforts of the damage control people who were working in the flooding engine room. The level of the water and oil had reached nineteen feet, and Lieutenant Hagstrom estimated it was coming in at a rate of four feet per hour. The men shoved the suction nozzle of one of their P-250 pumps through the oil into the water, where it would pump for several minutes before bobbing back up through the oil to the surface. As soon as it sucked in a glob of oil, now cooled and thickened by the cold sea water, the men were forced to shut it down, pull it out, and clean it. Then they pushed it back through the oil, only to start the process over again.

As they worked, the men noticed that the oil was inching its way up toward the main switchboard, which controlled the emergency generator. The hum of the generator was the sole source of power to the wounded tanker, the weakened heartbeat that kept her alive. If the ship lost its power, the damage control team would lose their lights and the ship's radio, which had provided good communications with the *Sherman* and the *Vigilant*. They would have only flashlights and walkie-talkie radios.

But the progress of the oil was relentless. And if it hit the switchboard, the men knew the board would short out, possibly in a flurry of sparks. The engine room was filled with fumes from pump motors and from oil. It was unlikely, they knew, but any spark from the switchboard could start a fire. To be safe, they would have to shut down the switchboard themselves before the oil could reach it. They informed the *Vigilant* and the *Sherman* of their plans, and at ten minutes to six they killed the power.

The generator sputtered, coughed, and went silent. The lights flickered for just an instant and went out. The ship was dead.

On the bridge of the *Vigilant*, several hundred yards away,

The Blackout

Ian Cruikshank gasped. He and the others with him knew it was going to happen. In fact, they had gathered on the bridge to watch. But they weren't prepared for it. The *Argo Merchant* had vanished. The blaze of lights against the pitch black sky had disappeared as if they had dropped into a crevice in the ocean. The crew of the *Vigilant,* stared into the blackness.

Cruikshank shook his head, half in awe, half in sadness. "That's it," he remarked to one of his officers. "You know, we're not going to make it with this guy."

Aboard the ship itself, it was dark and still. The damage control team had worked throughout the day with lights wherever they had needed them. Now, whether in the engine room or on the deck, they were in darkness. They had spent the day with the whir of pumps and the hum of the generator in their ears. Now, just one pump was running. For the first time, the people on board could hear the sound of the ocean working on the ship. Waves crashed against the hull, sending vibrations down the deck and an eerie rumbling echo through the engine room. They could hear the ship wrenching in the stress. Metal was creaking and groaning, as the weight of the water and oil in the engine room was slowly bending the hull. In the pitch-black engine room, the men shined their flashlights, making dull blotches of yellow on the gooey blanket of oil which was now four feet thick on top of the incoming water. As the ship rocked in the seas, oil sloshed up against a catwalk in the engine room on the level just below the main deck, pushing itself up through the iron grating like bread dough, then falling back as the wave passed under the stern.

The damage control team gathered in the mess area on the main deck in the afterhouse and took turns checking the one pump that remained working below them. When it failed again and bobbed to the surface of the oil after the lights went out, the men pulled it out for the last time. They couldn't keep it under water for long, and when they could, it was pumping a trickle while water and oil was coming in from somewhere else in torrents. Any further efforts to control the flooding were up to the Strike Team.

Whenever the men ventured out of the dining area now, they went in pairs, partly for their own safety and partly because none of them was interested in feeling his way around the vessel alone, amid the rocking, the noise, and the blackness.

6

The Condition

With the prime mover swinging easily beneath him as he hovered over the air station, Lieutenant McLean eased the throttle toward the fifty-five-knot maximum cruising speed. As he moved past thirty-five, the helicopter started shaking violently. The package beneath him had veered off uncontrollably, as if fighting the helicopter's course and wanting to fly on its own. McLean felt as if he was at the controls of a jackhammer instead of a helicopter. He eased off to thirty-five knots. The vibrations stopped. The pessimistic predictions for a successful mission were already proving out and he wasn't even over water yet. At the slower speed, McLean was now looking at a two-hour flight instead of what he had planned to be just over an hour. The additional time also meant that he would use more fuel. And he had already used up some of his meager reserve in his initial attempts to get the load off the ground.

While the pilots who had seen the ship during daylight had warned him about the cramped stern, they themselves had not viewed it as a major obstacle to their own operations. They could hover above the tanker at any comfortable altitude and lower gear and men down in the basket with a cable. If the basket was heading for trouble, they could stop the cable and move, or slow it down until someone on deck could grab it and guide it to a safe landing. McLean, however, had the 1,200-

pound prime mover on a twenty-five-foot tether. The height of the load itself added another five feet. He had no winch with which to raise and lower the package at his convenience. And no one on the deck was about to grab a swinging 1,200-pound payload and help ease it down.

"You got Chambers aboard?" he radioed back to Burnaw, who was airborne behind him with the Strike Team.

"Yeah, he's here."

"Ask him how much of a fall this thing can take. If I get it over the stern pretty well, two or three feet or so, I might just want to let it go."

"About six inches is all it will take, a foot at the most," Chambers, who was hooked into the conversation, informed him. "It won't take much of a fall."

"That's just wonderful," McLean remarked.

Once he found the tanker, he would have to swoop down into a hover and move the package at exactly thirty feet. Not twenty-eight feet. Not thirty-two. And he would have to be directly over a space on the deck that was clear enough to take the load. It would be like trying to thread a needle with a shoelace.

It wasn't until he was fifteen minutes away from the ship that he learned it had lost power. The lights were out. Now he had to thread that needle blindfolded.

"This is getting a little sticky," he radioed to Burnaw.

It was pitch black. He couldn't see the water. He couldn't see the horizon. And the "night sun," a powerful rescue spotlight that might have helped him, was on the other helicopter. He didn't even know if he would find the tanker.

But about 7:45, with help from the *Vigilant*, which was lit up nearby, he spotted it. From the air, it was nothing more than a few pinpricks of light from flashlights the men aboard were waving skyward, and a thin, dark gray outline of the stern just above the black ocean surface. He couldn't even see the obstructions on the stern, never mind distinguish a safe delivery area. Without lights, it would be impossible to put the Adapts aboard. He had two options, he thought. He could hover over the ship for a few minutes to see if the people on the deck could get some lights working on the stern. Or he could abort the mission and head for Nantucket. He didn't have the choice of returning to the air station: he was running out of fuel. And if he waited over the tanker too long, he would have to dump the prime mover if he were to make the island airport.

"There's no way I'm going to get it aboard that thing without lights," he told Burnaw. "My fuel is getting critical. I'm going to Nantucket. Why don't we all spend the night there and get this out there at daybreak. It's just impossible to put it on there now."

Burnaw agreed. He started changing the heading of the helicopter to a course for Nantucket.

Chambers, however, had other ideas. He had worked hard all day, and especially at the air station, to get the momentum of the incident working in his favor. Now, with McLean's decision not to attempt delivery of the prime mover, it was gone. It was replaced with what Chambers called a "condition," a negative train of thought, that he now had to overcome. But as a lieutenant commander, Chambers, like the pilots, was a lower echelon officer. He didn't have the rank or the authority to order any of them out there. He wasn't even running the show. He was working for Captain Veillette on the *Sherman* and Captain Hein in Boston. Besides, even if he were a captain, he couldn't order a pilot to fly a mission that the pilot considered too dangerous. Chambers had been confronted with these conditions in the past, and he had angered more than one captain, trying to take over the operation, before he learned that persuasion, not coercion, is his best way to get the momentum flowing his way again.

"Where're you going?" Chambers asked as he felt Burnaw turning the helicopter away from the direction of the tanker. Chambers was plugged into the radio circuit and he knew exactly there they were going.

"We're going to land on Nantucket," Burnaw replied.

"Why are we doing that?"

"The other pilot can't get your gear aboard. There's no light on the ship and he's running out of fuel. No sense in you sitting out there all night without the pump."

"Well, listen, is there any problem with going out there and just looking at it? I'd really like to see it."

Chambers' heart was pounding in his throat. Burnaw seemed like an easygoing guy, but if he had any reservations about flying any further he could stop the trip, just by saying no. Chambers' only alternative then would be to call Lieutenant Commander Pearce at the air station and try to force the issue. He didn't want to have to go that far.

"All right," the pilot said. "I got no problem with that."

Burnaw thought Chambers was damned adamant about checking the tanker out first hand. But it was going to be his baby, Burnaw said to himself, and the guy does need to know what's going on out there.

The pilot called the *Sherman* to inform Captain Veillette they would be out shortly to take a look at it.

"Do you think there's a chance we'll get the Adapts on there tonight?" Chambers asked.

"No. It doesn't look good," Burnaw replied. "That's going to be rough going in the daytime with that load underneath, never mind the nighttime when you can't see a thing."

"Well," Chambers said. "Let's look it over."

Looking it over was no easy task. They arrived at the scene just before 8 P.M., but they knew it was the scene only because they could see the *Vigilant.* They couldn't see the tanker anywhere. Chambers was amazed. He scanned the water below, only knowing it was water because they were, after all, supposed to be flying over the ocean. He didn't let on to the helicopter crew, but he knew now why McLean didn't want to put the Adapts package on the tanker. He could see almost as much with his eyes shut as he could with them open.

When they finally found it, after a fifteen-minute search, it was again with the help of the Coast Guard cutter and with the men aboard the tanker waving their weakened flashlights at the helicopter. When he saw the faint dots of light from five hundred feet, Chambers smiled. They sure aren't like the flashlights on the Eveready ad, he thought, the ones with the bright beams that shine into the darkness.

As they arrived overhead, Veillette called from the *Sher-*

man. Veillette thought the tanker should be evacuated. The pumps were all shut down, the people aboard had nothing to do, and the engine room was continuing to flood.

"The conditions are getting worse on there," Veillette said. "I'd like to get everyone off. The weather is supposed to pick up overnight and I don't want to lose anybody. But the master just won't get off. Do you think you could get aboard and try to persuade him to leave?"

"Yessir, I'll be happy to go aboard," Chambers said, "if the pilot here can get me down."

Chambers smiled broadly in the darkened cockpit. He had wondered how he would get aboard once he got this far, and he had found an unexpected ally: the on-scene commander.

He disconnected himself from the radio gear and stepped back into the cabin section of the helicopter, where Charles ("Mac") McKnight and Keith Darby of the Strike Team had ridden out with nothing but the noise of the engine and the clatter of the blades in their ears.

"I'm going aboard," he hollered to them.

Burnaw meanwhile made his approach to the ship. He switched on a large searchlight used to light up rescue scenes at sea. But he was too high for the beam to pick up the ship. The men on board the tanker shined a couple of small lanterns at the vessel's smokestack, and that was all the pilot could see. It was his only reference to the ship. As he descended toward the ship, he had his eyes glued to the illuminated stack to judge his altitude when the stack suddenly disappeared. In the beam of his own search light, he could see that he was flying through a cloud of smoke. He had lost sight of the ship and his perspective. And he was headed right for it.

Burnaw was startled for an instant. It was as if he had flown into a cloud he didn't know was there. Instead it was smoke that for some reason had belched from the *Argo Merchant*'s stack. But he was out of it again as quickly as he plunged into it. And he picked up the dimly lit stack again before he got into trouble.

Back in the cabin, Chambers and McKnight were leaning out the cargo hatch, trying to get a better look at the ship. Burnaw had the spotlight on it now and the beam was running along the catwalk, up the stack and around the stern. The superstructure cast long, eerie shadows on the aftersection of the ship as the beam searched out the best place to deliver Chambers.

"This looks like a piece of cake, boss," McKnight yelled over the din of the engine. They both grinned. Chambers thought it looked ghostly.

"I'll see you down there," Chambers said.

He crouched in the iron rescue basket and took a deep breath. He had stepped out of a helicopter like this dozens of times, but he had never gotten used to it. Earlier in his career, he had climbed 1,300-foot loran towers to install navigation lights, leaning off the tower on a strap. He had never gotten used to that either. But he made himself do it. It was part of the job.

The helicopter crewman pushed him out the cargo door, and he was lowered quickly toward the deck. The down draft from the helicopter rotor whipped pieces of the corrugated metal roofing from an awning on the fantail and a couple of men from the *Sherman* who had come out to meet Chambers struggled to hold them down. The basket bounced once off a metal brace and once off the awning before it clanked against the deck. Chambers pushed off on the sides of the basket and raised his body slightly as it hit so the base of his spine wouldn't bounce on the deck. He scrambled out.

He was greeted by Lieutenant Hagstrom, who ushered him into the crew's mess area at the stern where the rest of the damage control party were all sitting in the dark, chewing on biscuits and cheese. Hagstrom gave him a quick rundown on the conditions aboard: the engine room was flooding rapidly; their pumps were out of commission. Their flashlight batteries were running down.

"If you think your flashlights are low down here, you ought to see them from the helicopter," Chambers said with a smile.

He went to have a quick look at the vessel himself. He watched the flooding in the engine room for a minute, checked a couple of the cargo tanks for evidence that they had ruptured, and walked briefly about the deck to see how serious the starboard list was.

From the air, the ship had a ghostly appearance to it, and as badly as he had wanted to get aboard, Chambers was unnerved by it. But now that he had his feet on the deck, he was more comfortable. The ship was intact, he thought. His quick survey turned up no reason to abandon it, certainly not at this point. He asked two crewmen from the *Sherman* to take soundings around the ship to see just how she lay on the shoal. Then he got on his small portable walkie-talkie to Burnaw in the helicopter.

"It's not as bad down here as it looks from up there," he said. "Can you put my people down here?"

"Yeah, if you want 'em. We'll get 'em down there."

McKnight, a first class machinist technician, and Darby, a first class damage controlman, had both been on the Strike Team more than three years, and Chambers considered them the two best front line men he had—exactly the type he would want to be first aboard a grounded, darkened oil tanker in the middle of the night twenty-seven miles from the nearest shore. Both had two special qualities that Chambers looked for in every man on the Strike Team: they were optimists who didn't sit around wondering whether a problem could be solved. They assumed that it could, then set about solving it. And, once given an assignment, they each could carry it out completely, making their own decisions, doing their own improvising without further orders or supervision. In addition, Darby was an expert with a cutting torch, a talent Chambers had often put to use clearing metal obstructions to make way for hoses, pumps, and delivery of equipment. McKnight was a mechanic and Chambers' pump man. He knew the Adapts system like a GI knows his rifle—every piece of it from taking it apart, cleaning it, oiling it, testing it, running it, and treating it the way a teenager cares for his first car. With those two aboard, plus his own knowledge of the mechanics of ship salvage and his willingness to get his own hands dirty, Chambers figured he had all the expertise he needed to evaluate the problem they faced, set out a plan of attack, and get started before other members of the team could get aboard.

While Burnaw circled above the fantail, Chambers and the

Sherman crewmen tore the thin metal sheets of roofing off the shed in an effort to make more room for the basket. Then, one at a time, Darby and McKnight were put aboard.

"Welcome aboard, Charles," Chambers said, with more than a trace of affectation in his voice.

"This is a funny place to park your ship, commander," McKnight retorted.

Such formalities dispensed with, they settled down to business.

"Okay, Darb, you go find the cutting torch and clear the shit off that fantail. They'll never get our gear on here with the fantail in that condition. Mac, check out the engine room and figure out how we're going to rig the Adapts."

His men went to work. Now Chambers had to contend with his own optimistic assumption: that the gear would get there that night.

"What do you think about getting the Adapts on here?" he asked Burnaw, who was still circling the tanker, waiting for Chambers' decision on whether passengers would be evacuated.

"What, tonight?"

"Yeah."

Burnaw paused a moment and sighed.

"Well, where do you want it?"

"How about on the fantail?"

"Not a chance. It's too cluttered. Much too small and not enough light."

"Well, what about on the tank deck, between the two deck houses?"

Burnaw maneuvered the helicopter over the center of the ship. One of his crewmen, shining a small light on the area, noticed several antenna wires that spanned the distance between the two structures.

"We can't get in with those wires up there," Burnaw said.

Chambers had just noticed the wires himself. The ends were wrapped around cleats that Chambers could reach from the main deck, and he moved about unwinding the antennas and letting them fall. One of them, however, that was connected to the top of the smokestack wouldn't budge. He fought with the wire and the cleat without success. Exasperated, he disappeared into the engine room. Moments later, Burnaw saw Chambers poke his head out of the ship's smokestack. He had clambered and crawled up through the sooty flues and vents of the stack. He broke off the antenna with a hammer and dropped it to the deck.

"Come on in now and take a look," he said, still perched atop the stack.

Burnaw moved the helicopter in over the vessel's starboard side to take advantage of the wind which was blowing across the deck from the port side. He hovered directly over a spot where he thought the pump could be set down. About the same time, the H-52 from the *Sherman,* which was over to evacuate two of the *Vigilant'*s damage control people, hovered nearby, with his landing lights hitting the deck. Burnaw noticed that it provided a good reference for him and helped light up the delivery spot. The H-52 pilot, George Oakley, said he would help illuminate the deck if they made a second attempt with the pump.

As they sat in the hover over the tanker's midsection, Burnaw asked each of his two crewmen and his copilot, Bill Shorter, if they thought the task was possible. Each of them thought it was.

Burnaw didn't know what troubles McLean had confronted while flying the load out there. But, with the lighting from the Night Sun and the small helicopter, with the change in delivery point to the tank deck from the cluttered stern, and with the trial hover, he was convinced that delivery of the Adapts pump was possible.

"Okay," he said to Chambers. "We think it can be done. I'm not making any promises, but we'll give it another try."

While the assessment was their own, Burnaw and his crew had assumed they were actually gathering information for Rick McLean. They would brief him on what they had learned. Then he would refuel and make the attempt himself. But when they informed the air station they thought the package could be delivered, they learned that McLean had turned it down. He hadn't seen anything but a thin gray outline of the stern. He had wrestled with a vibrating aircraft at any speed over thirty-five knots, and he had nearly run out of fuel. Most important, he still hadn't seen the landing area, and it was still dark.

"If you think you can do it, why don't you try it yourself," Lieutenant Pearce suggested from the air station. "Take your helicopter to Nantucket, switch with McLean to get the one with the sling assembly attached."

Burnaw and Shorter had not anticipated those orders. Like McLean, they had the option to decline, but with their own optimistic evaluation, they were hard pressed to come up with an excuse not to try it. Besides, no matter what McLean had decided, it was logical for them to take it out there anyway, Shorter figured. They had already gone through a trial run.

Burnaw went around his crew one more time.

"I think we can at least give it a good shot, without getting in a bind," Shorter said.

The rest of the crew agreed. They headed for Nantucket to pick up the load. It was nearly 9 P.M.

"Strike Team One, this is the *Vigilant.*"
Chambers reached for his radio.
"Yeah, go ahead, *Vigilant.*"
"Hi, Barry. It's Ian."
Chambers smiled broadly.
"How're ya doin', Ian. Nice to hear your voice."

Ian Cruickshank was a good friend, and Chambers rarely had the luxury of working on a job with someone with whom he had an "across-the-bar-drinking-relationship." The Strike Team's mission is to fly in and out of strange places, where they always work with people they don't know under circumstances the local people may know little about. Chambers and the pilots had already sparred over the Adapts pump, creating a tension that might have been mitigated if they had known each other.

But Chambers would have no such battle with Cruickshank. They had known each other for the past four years, since Cruickshank had a desk job in Washington where he worked to set up a Coast Guard diving team. Chambers, who had just joined the Strike Team as its executive officer, second in command, was an avid diver and had met Cruickshank while trying to find good divers for the Strike Team. They had kept in touch ever since.

Cruickshank regarded Chambers as a rare type of Coast Guard officer who liked to get his hands dirty and who "struck first and worried about the consequences after." Most officers, he thought, measured their words more carefully than Chambers, who was an outspoken maverick with a penchant for rankling superior officers and with no patience for bureaucracy and inertia. But Chambers also had a universal reputation, even among those he had irritated, for getting the job done. If you need a man to come into a tough situation and take charge, Cruickshank thought, Chambers is your man.

"Look, I want to help you out any way I can," Cruickshank said, after giving Chambers a quick rundown on what had happened out there during the day. "If you need anything, just give a holler. I'll send it over on a helicopter."

It was a mental boost for Chambers. He not only had a friend who would look out for his needs on the tanker, but a man who understood his temperament and who could serve as a liaison to help fight his battles with the people on the beach.

Chambers also checked in with Captain Veillette on the *Sherman,* but he purposely waited until McKnight and Darby were aboard and the direction of events had swung back his way. He had gotten on the ship and he didn't want to give anyone a chance to order him off before he got his pump in operation. Veillette agreed that the Strike Team should stay aboard.

By 9:30, he got a report back from the *Sherman* crew on the soundings around the ship. It wasn't good news. The tanker was lightly aground in the forward section, where it had been bouncing off the bottom in heavy weather earlier in the day, and hard aground amidships, just behind the bridge house. The flooding engine room was hanging over fifty-four feet of water. If the stern continued to flood, Chambers thought, it would probably break off soon, and probably without warning. In thinking about how he would approach the stricken tanker before he actually saw it, Chambers knew instinctively that he would want his Adapts. Now the problem was staring him in the face. It was not a question of wanting the pump. He had to have it. If they couldn't get it out here before morning, they might have nothing left to bring it to.

Like McLean, Tom Burnaw and Bill Shorter had misgivings about the Adapts assignment when they first heard it under discussion at the air station that afternoon. By then, news reporters and scientists had begun to gather in the station and planes and helicopters were flying in and out like ambulances at a city hospital emergency entrance. The grounding of the *Argo Merchant* was already an event with its own momentum. It placed another dimension of pressure on the participants in the event, pressure that Burnaw and Shorter thought unquestionably contributed to the air station's decision to take on the Adapts delivery. The pilots were wary of being swept up in the event, unable to make a hard, cold kind of cost-benefit analysis of the mission, especially since the cost could be human life. Chambers' relentless prodding was a factor as well. The pilots considered him a likable, forward character and something of a daredevil. He was probably perfect for the Strike Team, they figured. But a daredevil helicopter pilot is one of two things: he's lucky or he's dead.

If the ship split, and its oil polluted the coastline, it would be a shame, Shorter thought. If he could, he wanted to help prevent that from happening. But he didn't want anyone to die in the process.

Burnaw was particularly concerned about doing the job at night. Helicopter pilots need what they call "visual reference" —a stable bench mark they can watch while they hover or fly so they know where they are relative to the ground or sea below them. Flying over water at night, a pilot has no way of distinguishing between the sea and the sky. Except for the glow of lights on his instrument panel, he can see nothing until he is close enough to the water to see the reflection of his landing lights on the surface. Most helicopter accidents at sea occur, Burnaw knew, when pilots without a visual reference lose their depth perception, misjudge their altitude, and fly right into the water. To get the Adapts onto the *Argo Merchant*, he would have to descend from his flying altitude toward an ocean surface he could not see over a ship he could not see. And with the Adapts package swinging from his underbelly, he had a load he could not see. He and Shorter would have to rely on their crewmen to guide them into a position to release the load.

When they landed at Nantucket, McLean briefed the two pilots on the difficulties he had with the sling load. They decided to combat the vibration problem by attaching a steadying line to the package and run it up to the cabin where one of the crewmen would keep it taut. They detached the Night Sun from their helicopter and put it on the one they would take to the tanker. They put in less fuel than McLean had needed since they were flying the package over a much shorter distance. And McLean told them of the sticky electrical release button. He had had trouble with it again on Nantucket.

Burnaw and Shorter lifted off the runway about 9:45, took the load up to seventy-five feet, and hovered for a few minutes, testing out their power and getting a feel for the weight beneath them. It felt okay. Still, Burnaw had no illusions that the trip would be easy.

"If this thing turns to worms, I'm not going to hesitate a minute," he told Shorter. "I'm getting rid of it. I have no qualms at all. It's been push, push, push to get this thing out there, and they'll really be ticked, but that's the way it goes. If we're in trouble, that thing gets jettisoned."

He also told Shorter that when it came time to release the load on the deck, he would try the electrical release once. If it failed, he would pass control of the helicopter to him and kick the foot plunger.

On the way out, Mike Ryan, one of the crewmen, strapped himself into a leather harness, grabbed the steadying line, lay down on his stomach, and leaned nearly half his body out the cargo door so he could tend the load. The other crewman, John Fagan, held a light on the package so Ryan could see it. Ryan filed an almost constant report to the pilots on how the load was riding, whether it was steady, starting to swing, or starting to rotate. With the added directions and the tending line, Burnaw was able to fly at fifty knots, about the speed McLean had hoped to make on the trip from the air station.

As they neared the tanker, the pilots radioed the *Sherman* and the *Vigilant* to get a wind direction. While the wind was light, they wanted to be sure they had every advantage available to them. The two Coast Guard cutters, however, each gave different wind conditions, and both were different from the wind they had on the first trip out. On his radio, Burnaw heard Oakley above the tanker in the H-52 and asked him to drop a smoke flare when they got out there so they would have an accurate wind line.

By 10:20, they were over the tanker. Burnaw began a slow descent, his eyes riveted to his instrument panel, especially his altimeter, which measured his distance from the water. Shorter looked out the window, searching for the tanker and keeping his eye on the H-52, which was already hovering off the deck. Instead of flying at a shadowy outline of a hull, they could use the other helicopter as a reference. They settled into a hover off the tanker's bow, about fifty feet above the water.

Burnaw had Oakley and the two Coast Guard cutters on the radio. Chambers was listening in from the tanker's deck.

"Okay, I want to make one shot, one nice straight shot to the ship and let it go," Burnaw said. "I don't want to be fooling with this thing. There's a chance it could get out of hand and I'll have to get rid of it. I want to get the wind line, make one shot right into the ship and get the damn thing over with. Okay, George, drop me a smoke."

Oakley dropped an eighteen-inch salt-activated flare from the helicopter. It burst into a bright pink fluorescent flame as it hit the water, and gave off enough light to illuminate a thin line of smoke trailing off in the breeze. The wind was blowing across the deck, from starboard to port, 180 degrees from its direction on their first trip out. The plan had to be changed. They would deliver the package to the port side of the deck.

Burnaw scanned that side as he prepared to make the approach. Chambers had moved over there to await the prime mover. Shorter ran through standard instructions and stressed that no one touch the load until it was on the deck and was released from the helicopter. The last thing they needed was someone strapping the load down before it was unhooked. If the release failed, then the helicopter would be tied to the tanker.

Burnaw looked at the H-52, just off the starboard side, shining its light on the deck.

"Okay, George, I'm going in from the port side. I don't want you in front of me. Just slide around in behind me and get your lights on it from there."

The small helicopter moved off to Burnaw's right, its beam falling directly on the delivery point.

"Okay. Here we go."

From the hover off the bow, Burnaw guided the aircraft down the port side of the ship and eased it in over the deck between the bridge and the afterhouse. Gusts of wind from the rotor swirled up off the deck, jostling the payload. It started to twist on its tether.

"It's spinning," Ryan said, pulling on his tending strap trying to steady it.

The helicopter started to vibrate. Burnaw fought with the controls, trying to keep it steady. Ryan barked directions.

"Easy does it. A little forward."

The vibrations increased. Burnaw felt the tail of the helicopter start to gyrate.

"Now to the right. She's bouncing around. Steady now."

Burnaw's grip tightened on the controls as if he were trying to crush them. The helicopter shook almost uncontrollably.

"Okay, lower it down. Easy. Down a little more. It's on the deck. Give it slack. Easy. Okay. Release!"

Burnaw pressed the button. Nothing happened.

The jerking of the aircraft would pick it up again if he didn't move fast. On the deck, Chambers clenched his fists. He didn't move.

"I'm passing the controls to you, Bill."

For a delicate second, both men were flying the helicopter. Then Burnaw let go. Shorter gripped the controls. The helicopter was bucking.

Burnaw grabbed the sides of his seat. He raised his body off

the cushion and tromped on the plunger. The hook let go.
"Its clear!" Ryan shouted. "We're released!"
Shorter urgently worked the controls. The helicopter steadied itself and rose abruptly above the deck.
Chambers exhaled deeply, glanced at the prime mover, and then looked up at the helicopter.
"My friend, that was beautiful," he said. "Just fucking beautiful."

7

The Rigging

As they turned away from the *Argo Merchant* and headed back to the Cape Cod Air Station, Burnaw and Shorter turned toward each other in the darkened cockpit and laughed.

"What the hell are we doing out here?" Shorter wondered. "What are we doing sling-loading equipment in the middle of the night when we haven't even seen it in daylight?"

They both shrugged.

"I'm just damn glad it's over," Burnaw remarked. "I still don't understand why they had to have it tonight. I hope it's worth it to them."

They radioed the air station with word that their mission was accomplished and they were coming home.

"Okay, we've got another load for you," the response came back. "We'd like to get another load out there."

Their faces went blank. The euphoric feeling of accomplishment and relief they had just begun to savor drained from both of them.

"Another load?" Burnaw asked incredulously.

"Yeah. We've still got the pump here on the ground, and all the gear with it."

Being unfamiliar with the Adapts kit, the pilots hadn't even thought about the second pallet. The prime mover they had just put on the deck of the tanker was a critical piece of equipment:

it ran the pump. But it was only a diesel engine. Without a pump, not a drop of water or oil would get sucked out of the engine room. The prime mover would be useless. And the effort to get it there would be nothing more than a training exercise. But the pilots felt deceived. It was the prime mover that just had to get out there, Burnaw thought. That's what everybody was pushing. Okay, it's there. But no one had said anything about more.

"Aren't we pressing our luck a little bit?" Burnaw asked the station. "With that load swinging underneath, it was like a herd of buffaloes walking around inside the helicopter."

But the officer at the air station had a bargain: the second load could be broken down and loaded piece by piece inside the helicopter. The pilots wouldn't have to carry out another load underneath them. They could hover over the tanker and lower it on the hoist.

Burnaw looked at Shorter and shook his head. They had little choice. Their night wasn't over yet.

"Yeah, okay," Burnaw said. "As long as we don't have to sling load it out here. We'll do it."

Back in Boston, Captain Hein was already making arrangements to get two more Adapts systems aboard the *Argo Merchant*, not by helicopter, but on the *Bittersweet*, under the plan originally conceived by the air station. Officials at the air station weren't interested in airlifting any more Adapts kits out there, day or night, as long as a ship was available. The weather forecast for the next twenty-four hours ruled out flights anyway. They would have to rely on the buoytender.

The plan was to load the pumps aboard flatbed trucks and transport them to Woods Hole during the night and lift them aboard the *Bittersweet* in time for a 6:30 A.M. sailing. That would get them out there by early afternoon. Hein expected Yokohama fenders would be at the air station by then and could be shipped to the tanker on the same trip.

Hein also got word from New York that a lawyer and salvage officials representing the owners of the vessel were due on Cape Cod by 9 P.M. and wanted to get aboard the tanker as soon as possible. The captain arranged for a flight from the air station.

Meanwhile, the tug and barge that left Sandwich shortly before 6 P.M. had headed toward Provincetown to make the trip around Cape Cod. It would have been shorter to cut through the Cape Cod Canal to Buzzards Bay and sail around the Cape Cod islands, but the tide in the canal was running against the barge and, because of its size, it was not permitted to struggle through against it.

Unknown to the Coast Guard, however, the tug and barge were beaten back by high winds and heavy seas as they rounded the tip of the cape and the skipper sought refuge in Cape Cod Bay, just off Provincetown, where he dropped anchor. He would not be at the scene by dawn.

In addition to the Yokohama fenders, the huge barge would need good weather before it could move in close enough to the tanker to receive its oil. The maximum conditions it could withstand were twenty-knot winds and four-foot seas. Despite the rough weather off Provincetown that night, the sea conditions at the scene of the grounding were within those limits. But the National Weather Service, which was issuing special advisories for the Nantucket Shoals area at the request of the regional response team, predicted weather there would worsen the next day.

After the prime mover was put aboard, Chambers made a more thorough tour of the tanker and spent time in the engine room watching the flooding. Water and oil was a little more than twenty feet deep when they had first come aboard, and it had increased about four more feet since then. The engine room was just over half full, and at 11 P.M. the level was rising at about eighteen inches an hour.

McKnight already had figured that the Adapts pump would have to be suspended from a hoist at the top of the engine room and lowered down through the oil into the water. Cruikshank had sent over a clamp, which they had screwed to a beam, ready to be rigged with the hoist when it arrived with the second load of gear.

As Chambers inspected the deck on the listing starboard side, he looked over the rail to see just how far down the hull had settled. What he saw was disquieting. The portholes, which ran just below main deck level along the stern section, were nearly in the water. And they were open.

If the portholes went under, water would quickly flood the crew's quarters, flow out into the passageway around the stern, and pour into the engine room. Water would not only be flooding from the bottom; it would be coming over the top. They would lose the engine room for sure, he thought, and they might lose the ship as well.

He hurried over to the port side, where McKnight and Darby were testing out the prime mover.

"Darb, Mac. We gotta close those portholes over on the other side. They're stuck open and they're damn near underwater."

They recruited the four remaining members of the damage control party from the *Sherman* and headed for the passageway to the rooms where the portholes were located. The crew's living quarters ran along the starboard and port sides of the stern section and were connected at the stern by a lounging area and mess deck where the crew ate, and where Chambers had found the damage control party waiting for him when he came aboard.

When Chambers and the others turned the knobs on the doors to the crew's rooms, however, they couldn't get in. The doors were locked. The men slammed their hips and shoulders against them, but the doors were heavy, thick steel. They didn't break easily. When the crew had been forced to leave the ship virtually empty-handed, they had stowed their belongings in their rooms and locked the doors.

"They got some crowbars and fire axes and stuff in one of the holds below," Darby said.

Darby had already inspected most of the storage lockers when Chambers sent him off to find the cutting torch. During the search he had routinely checked for tools, rope, and other gear that the team might be able to use. He had found the cutting torch, but his work clearing some of the metal off the fantail turned out unproductive when Chambers and Burnaw had decided the prime mover would be delivered to the tank deck amidships. His quick inventory of the other equipment aboard, however, was about to pay off. The men got the axes and crowbars and bashed their way into the rooms. Darby cut locks off a couple of doors with the torch.

As they got the doors open, the men, holding flashlights, made their way carefully through the dark rooms along a slight decline in the floor to the side of the ship. Water had already splashed through the portholes and was sloshing around almost knee deep in the low end of the rooms where the men had to stand to work. Their boots filled up with cold sea water.

When they tried to close the portholes however, they wouldn't budge. The hinged levers that held them open and the screws and fittings that helped batten them were smothered with paint.

"Shit, these haven't been closed in years," Chambers said as he shined his flashlight on the bulging layers of paint. With the engine room close by, he thought, the rooms must have been hot all the time, so the portholes were kept open for ventilation.

To close them the men had to pry up the hinged levers, which the paint had fused to the porthole form. With crowbars and pieces of pipe, the men braced themselves against the side of the ship in the darkness and pried and jimmied and slammed and bashed the levers. Chambers worked in one of the heads, standing on the toilet seat to get good leverage on the porthole with a crowbar. Waves occasionally blew through the porthole, showering him with salt water as he worked. More than once, his foot slipped off the wet sides of the toilet into the bowl. He spent ninety minutes trying to close that one porthole, and finally succeeded when Darby came in and they both yanked on the crowbar. The other men, Chambers knew, were having the same trouble. He could hear McKnight through the bulkhead in a nearby room swearing loudly and clubbing the porthole with a pipe.

By 1 A.M., they had done what they could with the two dozen portholes on the starboard side. All were closed, but none was watertight. As they got the last one shut, the water level along the hull had reached the lower rim of the portholes. Now, however, the sea would only trickle through cracks instead of pouring into the rooms.

The men also worked to close the port side portholes. That was a contingency. The engine room could sink, Chambers thought, especially if the pump didn't get out there. To refloat it, they would have to pump air into the holds. The portholes would have to be closed to make the rooms as close to airtight as possible. It was better to get them shut now, while they waited for the pump, Chambers thought, than to send divers down to close them later.

By 1 A.M., Chambers also wondered where the pump was. Closing the portholes was just a stopgap. It wouldn't hold the seas back forever. And the pilots had delivered the prime mover nearly three hours ago. To Chambers it seemed like twenty-three. He got on the radio to Cruikshank.

"Tell them to get off their asses in there, will ya? I've gotta have that pump out here."

He stalked about the ship. He checked the prime mover. He checked the engine room. He checked the tanker's list. And he looked skyward, hoping to hear the thumping sound of a helicopter. The level in the engine room was approaching thirty feet. The list was increasing steadily, and by 1:30 some of the portholes on the starboard side were completely underwater.

As he walked out the passageway on the port side from the engine room toward the tank deck Chambers nearly stumbled over a distorted mound of metal that seemed to have grown out of the deck. He had been down that passageway several other times that night, and he had not seen the hump before. The deck had started to buckle.

From the moment he stepped aboard, Chambers had known the weight in the engine room was pulling the stern down into the water, bending the hull about midway between the bridge and the afterhouse. But now something else was happening as well. The constant side-to-side rocking of the tons of water and oil in the engine room had finally started to twist the deck just above it. The engine room couldn't take much more. Chambers knew it would not be long before the water on the starboard side was over the main deck. Then, even the best efforts couldn't keep it out of the engine room for long. Chambers grabbed his radio again.

"Look, Ian, I got water coming up to the main deck. The engine room is starting to twist. And my fucking gear isn't here yet. Where the hell are they? Make sure they understand this: if I don't have that pump here in an hour, I'm going to lose this rig."

Then he ordered everyone off the stern. He got Darby and McKnight out of the engine room. He made sure the damage control people cleared out of the mess area. And he did a room inspection. One of the eight remaining crewmen had curled up in a blanket on a cot in the lounge area, where Chambers routed him out and sent him to the forward lounge area, a large salon two decks below the bridge. Chambers wanted everyone in the same place and on the high side of the ship if the engine room broke off.

Just after 2 A.M., he heard the faint chopping sound of a helicopter overhead. His equipment had arrived. But the anxiety remained. Chambers didn't know whether it had arrived in time.

Of the three and one half hours that had elapsed since Burnaw and Shorter had put the prime mover aboard, the pilots had spent nearly half in the air, flying to the air station, a seventy-five-mile trip, and back to the tanker again. When they landed at the air station, other personnel started unpacking the second Adapts pallet and loading it and some other gear onto two helicopters. The total weight of the second load would run to 2,200 pounds, and when air station officials saw that, they decided a second aircraft was needed. Otherwise, pilots would have to make a third run.

While the aircraft were being loaded, Burnaw and Shorter had a couple of cups of coffee and briefed the pilot of the second helicopter on flying conditions out to and around the tanker. They also recalculated their fuel-cargo weight ratios to determine how much fuel they could carry. Fourteen hundred pounds of equipment was loaded into one helicopter, about the total weight of the prime mover, and another eight hundred pounds was put aboard the second, which Burnaw and Shorter would fly. Four passengers, the representatives of the tanker owners, also were to go in the second aircraft.

It was just after 1 A.M. before the helicopters were loaded and ready to go. The pilots climbed into the cockpits, tested their engines, and by 1:20 both were airborne for the forty-five-minute flight back out to the *Argo Merchant*.

As the two helicopters approached the tanker, Burnaw and Shorter held back. They had the Night Sun, so they hovered off to the side and shined the light on the deck for the crewmen in the first helicopter. Then, slowly, the piece-by-piece process of lowering the rest of the Adapts gear to the deck began.

When the 450-pound pump hit the deck, Darby and a crewman from the *Sherman* each grabbed an end and dragged it down the deck to the doorway into the afterhouse. They lifted it up over a six-inch lip into the dark, narrow passageway where Chambers had noticed the buckling. After sliding it around the distorted hump, they dragged it down to another doorway, lifted it over another ridge, and shoved it across the vessel onto the catwalk on the starboard side of the engine room, finally guiding it to a point under the beam clamp on the ceiling. McKnight and Darby had rigged the clamp on the starboard side over a space that dropped clear to the bilges, avoiding the tangle of catwalks and decking that resembled scaffolding as well as the turbines and boilers and other parts of the ship's engine.

Then McKnight and the *Sherman* crewmen shuttled equipment back and forth from the delivery area on the deck, taking shackles, straps, harnesses, and a hoist that were needed to suspend the pump from the ceiling. Darby did the rigging.

The six-inch diameter discharge hoses, weighing 185 pounds and rolled up as big as truck tires, and the hydraulic hoses, which carried the power to the pump, had to be laid out in the dark passageways and hooked together. As Lieutenant

Hagstrom helped McKnight couple sections of hose, he paused for a moment, smiled, and looked carefully at the man on the other end, wondering just a little about who should be working for whom.

"By the way, what's your rate?" he asked.

"Well, sir, what do you think it is?" McKnight replied with a grin.

"I'd say it's either first class or chief, and I'm not too sure I like the idea of working for a first class."

"Well, how do you feel about a chief, sir?"

"I don't really relish working for one of them either."

They both laughed. McKnight suppressed a smirk. He was just a first class, one notch under a chief and several notches under a lieutenant.

"Well, just a couple more minutes, sir, and we'll be through," he said. "Then you can relax and get some sleep."

All the *Sherman* people were exhausted, McKnight thought. They had been on the ship all day working almost without stopping. They were dragging. But McKnight was glad they had stayed aboard. Without their help getting the equipment laid out and hauled to the proper places, the three Strike Team members would have had to do it alone, adding precious time to the task. No one knew just about how much time they had, but the signs were not good. As the men worked in the passageways, bulkheads made of a hard wallboard material started to crack. And one split open from floor to ceiling, exposing the steel frame of the ship behind it, as the vessel writhed in the strain of the weight on her stern.

By the time they finished rigging the hoses, the passageways were clogged like the streets at a general alarm fire. The helicopters completed the delivery of Adapts gear by 3:30 A.M. and returned to the air station for the final time that night. Within a half hour, the final connections on the pump were made. The discharge hose ran out an open porthole on the port side.

Darby, who was stationed in the engine room, lowered the pump with the hoist down into the goo. It slipped easily through the oil and disappeared. Only the six-inch yellow-brown discharge hose emerged from the black liquid, like a jungle vine growing out of a swamp. The pump was ready.

The portholes on the starboard side were now three feet under water, and the passageway from the tank deck past the crew's quarters to the engine room had begun to fill with water. Chambers, who was watching the pump sink into the hold, noticed with a start that the door at the end of the passageway out to the deck was still open. He dashed down the corridor and looked outside. The sea was within six inches of pouring over the lip of the doorway.

He closed the door and screwed it shut. He paused a moment to shake his head, wondering how they had forgotten to close it. Six more inches and that water would have cascaded into the engine room, he thought. They would have lost it before they even got their pump going.

He made his way back down the corridor and over to the port side, where he went back out on deck to watch the end of the discharge hose while McKnight started up the pump.

"Okay, Darb, you ready to go?" McKnight shouted into his radio as he stood next to the prime mover, a few feet away from Chambers.

"Yup. Start crankin' her up, Mac."

Darby was in the engine room about seventy feet away. He shined his flashlight on the limp, rumpled hose that protruded from the oil a few feet below him.

McKnight eased the prime mover into gear and turned up the pressure.

"Okay, Mac, I've got pressure coming through the hoses," Darby said, watching the hydraulic lines that brought the power to the pump. Then he listened. The pump sucked water into its head and propelled it into the discharge hose.

The water pushed its way through, inflating the hose firm and perfectly round as it surged out of the engine room.

"You got discharge yet, Mr. Chambers?" he asked.

"Nothin' yet."

"It should be there any minute."

Chambers leaned over the rail. The limp end of the hose suddenly whipped itself hard against the side of the porthole. Air rushed out the end with the shallow whoosh. Water gushed into the ocean.

"We've got discharge!" he hollered. "We've got discharge!"

8

The Progress

For the first time in a night filled with frustration, Chambers allowed himself a smile as he watched the water from the engine room pour overboard. The Strike Team had coined an expression for just such moments: "Happiness is a hard hose." Chambers, McKnight, and Darby went through the proper ritual of repeating the phrase to each other and laughing, as if it were a joke they had never heard before. None of them knew yet what effect the Adapts would have on the flooding, but, for the moment, it was enough that the pump was working.

Chambers gets satisfaction from his ability to keep oil from spilling into the ocean, but he savors at least as much the feeling of accomplishment he gets just from being able to move and set up the oil spill equipment quickly and get it running. Though few jobs have a more direct impact on the environment than fighting oil spills, Chambers does not consider himself an avid environmentalist. In fact, he shuns the label. He is a technician. His wife Cindy once gave him a sign that hangs in his office in Elizabeth City and reads: "The difference between men and boys is the price of their toys." The Strike Team has some expensive toys, and while the men get satisfaction from using them to protect the environment, they also particularly enjoy just making their toys work.

Within the first hour, Chambers, McKnight, and Darby had

evidence that the Adapts was working just fine. The level of the liquid in the engine room had dropped nearly a foot. The pump had to be shut down once to remove a burlap bag that got sucked into it, and once to reroute the discharge hose out onto the main deck so it wouldn't slip out of the porthole and pour the water right back into the ship. Otherwise, the pumping went smoothly.

At 6 A.M., while Darby took a turn watching the prime mover on the deck, Chambers and McKnight sat on a railing along an oil-smeared platform one deck down in the engine room, smoking cigars. The beam from their flashlights and the red glow from the cigars were all that lit the space.

McKnight, a stocky man with a round face and a crewcut, was only a first class, and several other Strike Team members were superior to him in rank, but Chambers considered him the team's mainstay. Both men were divers and had the kind of camaraderie and trust in each other that comes from working together on difficult and dangerous underwater assignments. They had met in 1972, in Miami, where McKnight was doing security dives, checking vessel hulls for bombs and other weapons which might upset the order of the carefully planned Richard M. Nixon Republican Convention. Chambers was there for a routine Coast Guard diving requalification, and the two of them had spent a day diving together. A few months later, when Chambers was trying to establish a group of divers on the fledgling Strike Team, he had called New Orleans, where McKnight was a diver in a special marine safety inspection program.

"Hi, Mac," Chambers had said, "How much clean underwear you got?"

"I'm *not* coming!" McKnight had replied without a second's hesitation. McKnight knew the obscure underwear reference was to the frequent travel assignments he would get as a member of the Strike Team. That didn't bother him much, but he had heard the team was to be based in New York, and he didn't want any part of that city.

"We're not going to be at New York," Chambers had said. "They're putting us down at Elizabeth City."

"Well, no, I don't think so."

"Think it over. We want some good divers on this team. I'll call you back."

"Well, all right, I'm not too keen on the idea, but I'll think about it."

"You've thought about it long enough. Pack your bags."

McKnight had paused, shaken his head, and sighed heavily.

"Well, all right. I'll take the orders."

Chambers still had to persuade Coast Guard officials in Washington and New Orleans to release McKnight from his diving assignment, and McKnight knew some of those officials were reluctant to do that. But Chambers pulled it off. Now, thanks to that one phone call, instead of inspecting vessel hulls in the calm, warm Gulf Coast waters around New Orleans, McKnight was deep in the bowels of a dark, disabled tanker in the frigid waters off New England, listening to the hum of the Adapts and arguing with Chambers about how many ladder rungs running up the side of the engine room had reappeared since they started the pump.

To settle the dispute, Chambers started marking the pump's progress by drawing a line with his finger every ten minutes or so on an oil-coated ladder right at the level of the liquid. They could watch it go down, slowly but steadily, as the pump sucked the water out. Chambers sat there in the dark with the glow of his cigar illuminating a confident grin on his face as he played the beam of his flashlight high on the bulkheads above him.

When the beam of his flashlight fell on the forward bulkhead, which separated the pump room from the engine room, Chambers saw what looked like moisture up high where the bulkhead should be dry.

"Hey, Mac, put your light up there. What's that look like to you?"

With the shine of both lights, they could see it clearly: oil was leaking through a couple of rivets about twenty feet over their heads.

As they looked closer, Chambers noticed that the bulkhead was flexing back and forth, as if it were breathing. His stomach tightened. If oil was behind that bulkhead, it could collapse any time.

"Hmm," he said, looking at Mac. "There isn't supposed to be oil up there."

They were quiet for a moment. They could hear the oil moving behind the bulkhead as the ship rocked in the swells.

"Well, boss," McKnight said, "I don't think I want to stay down at this level too much longer."

On that, there was no argument. The two of them scrambled up the ladder. McKnight stayed in the engine room, at the main deck level, to keep his eye on the Adapts, while Chambers went to see what the pump room looked like. He opened the same manhole in the deck that Ypsilantis and the damage control team had opened earlier in the day, and looked in. It was full of oil.

"Shit, I looked in there when we came aboard," he told McKnight. "It was just a quick look, but I didn't see anything like that in it."

The oil in it now, Chambers thought, explained the oil in the engine room. The pumps are powered by motors in the engine room which are connected by shafts that run into the pump room through holes in the bulkhead. The holes are supposed to be sealed to prevent oil fumes from escaping into the engine room, but Chambers knew it was not unusual on tankers for the seals to be broken or missing entirely. Thus any abnormal build-up of oil in the pump room would eventually flow through these holes into the engine room. The question was, how did the oil get into the pump room in the first place?

Oil in a pump room was nothing new to Chambers. He had frequently found it floating around on foreign-registered tankers in places where it shouldn't be. But that much cargo behind that bulkhead, he thought, meant one of two things: either the bulkhead separating the No. 10 row of tanks from the pump room had caved in, or the tanker's internal plumbing system, which regulated the flow of oil in and out of her thirty tanks, had ruptured. Since he hadn't seen any indication that the structure of the ship had weakened sufficiently to cause a bulkhead to collapse, Chambers figured ruptured piping was the likely explanation. And that was the worst of the two possibilities. If the piping damage were particularly severe, it meant that he could have the ship's entire cargo backing up into the pump room, against the bulkhead, through the pump shaft holes and into the engine room. And as the Adapts continued to move the water, it was approaching the time when it would have nothing left to pump but oil. Chambers didn't have any place to put oil yet. Besides, the gunk already in the engine room had congealed when it hit the forty-degree sea water. By the time it floated to the surface, it was like paste. He and McKnight had tossed a claw hammer into the goo and had watched as it took thirty seconds to sink, slowly oozing into the oil like a man's hand disappearing into quicksand on one of the old grade B jungle movies. If too much of the oil got that cold, they would have to find a way to heat it before they could pump any off. If they were unable to pump it, the oil would flow relentlessly until the levels in the engine room and in whatever tanks were feeding the engine room were equal.

The pipe rupture theory also helped explain to Chambers how water was getting into the engine room. Since the Adapts pump was making good progress, moving water out at about one thousand gallons a minute, water had to be coming in slowly. It wouldn't take much of a hole for the flooding to overtake the rate of the Adapts discharge, Chambers thought. But if a pipe carrying cooling water for the boilers had broken, for example, the water might be coming in through a four-inch hole. That would account for a 200- to 300-gallons-a-minute flooding rate, he determined, and explain why the Adapts was so successful. Of one thing Chambers was sure: water was not coming in through a hole in the hull of the tanker under the engine room. That had to be sound there, he thought. After all, it was hanging over fifty-four feet of water. The ship had stopped before the engine room could have skidded over the shoal.

Cmdr. Dominick Calicchio from Captain Hein's office, who had come aboard with Chambers and the other Strike Team members, was also trying to determine why the pump room was full of oil and how oil and water were getting into the engine room. While he had taken time to help the Strike Team close the portholes, Calicchio's primary assignment was to evaluate the tanker's condition and report back to Hein and other officials ashore.

As he toured the ship, he found that the bridge was neatly kept, but he thought much of the rest of the vessel was poorly maintained. Rust had accumulated on ladders, deck valves, and tank tops. Ladder rungs had rusted through. Rubber hoses ran out of at least two tanks in the forward section of the ship. The crew had told him the hoses had carried steam into the tanks to keep the cargo warm on the voyage up the coast. The heating coils in those tanks were broken, the crew had explained.

Calicchio also found heavy layers of rust on many reach rods, which crewmen turned on the deck to open and close valves at the bottom of cargo tanks. A tanker's plumbing system is organized so every tank can be isolated from the interior cargo pipes and from every other tank. Thus, if a cargo pipe had ruptured, the ship's crew theoretically could have sealed off all the cargo tanks, and only the oil in the pipe itself could have flowed into the pump room. The engineer had told Calicchio that all the valves, including those operated with the reach rods, had been closed as soon as the ship had run aground. But the pump room had much more oil in it than could have come from one cargo pipe, Calicchio thought. And he tried without success to turn several of the reach rods. None of the rust he saw had fresh scratches on it or had been broken and knocked away. Despite the engineer's insistence that the valves had just been closed, Calicchio didn't think they had been turned in months. It was possible they had jammed when the vessel stranded, but he considered that unlikely. The rust as well as the oil in the pump room, he thought, proved his case.

The tug and barge hired by Captain Hein left the refuge off the tip of Provincetown at 6 A.M. Rather than test the rough, open ocean around Cape Cod, the skipper of the tug chose to turn back through the sheltered waters of Cape Cod Bay, the canal, and Buzzards Bay enroute to the tanker.

Neither Captain Hein in Boston nor Captain Veillette on the *Sherman* knew, however, that the tug had stopped in the first place. They were still expecting the barge on the scene about dawn. Veillette even sent a message urging that the Yokohama fenders be transported out there quickly because the weather had begun to worsen shortly before daybreak and he thought it important to get the fenders and barge in place by the tanker before it got much rougher.

The *Bittersweet* was at Woods Hole, loaded with two more Adapts pumps for Chambers, and ready to sail. But her skipper was waiting for the fenders. Hein was counting on the *Bittersweet* to take them to the tanker. He had no other means of getting two 5,000-pound pieces of equipment out there. But after catching a couple hours' sleep in a bunk at First District headquarters, he awoke to find that mechanical problems with a C-130 had delayed departure of the fenders from Elizabeth City. They wouldn't arrive at the air station until 10:30 A.M. and they would still have to be trucked to Woods Hole for loading onto the buoytender. The Adapts pumps couldn't wait that long. But with the news that the tug and barge had been delayed as well, it made no sense to wait for the fenders anyway. The *Bittersweet* left Woods Hole shortly after 8:15 A.M.

Hein now had to find a new way to get the fenders out there. They were much too heavy for the air station's H-3 helicopters. The Army representative on the regional response team had an answer. With a couple of telephone calls, he could get a helicopter for Hein that could pick up both fenders at once. It was known as a Skycrane. By 11:30 A.M., word went out to Fort Eustis, Virginia, to fly two Skycranes up to Cape Cod by 9 P.M. that night.

Meanwhile, Hein hired two more tugboats, the *Sheila Moran*, and the *Moira Moran*, both out of New York. The *Sheila Moran* had no barge, but Hein thought it could be used to tow the tanker, once pumping operations had lightened her of enough water and oil to make a reasonable attempt to pull her from the shoal. It was due at the scene by late afternoon. The second tug, with a 40,000-barrel barge, was expected the next day. Hein and other officials ashore also activated the Gulf

Strike Team, ordering five men and one of their Adapts units flown in from their headquarters in Bay St. Louis, Mississippi. And, bracing for a spill, Hein requested three boxes of high seas oil containment boom—more than 1,800 feet—from the Atlantic Strike Team in Elizabeth City.

The mobilization of equipment and people was continuing as if the Coast Guard had taken over the ship. Indeed, with the Strike Team aboard, it was in control of the ship. But the First District had not yet made it official by invoking the Intervention Convention. Admiral Stewart preferred instead to give the owners a few more hours to get some salvage equipment at the scene themselves. Captain Hein called Amership in New York that morning to check their progress and to explain the convention to them.

A salvage officer and three other company agents were already aboard, having gone out on a helicopter with the second Adapts delivery at two in the morning. Two more men were on the way to Cape Cod. And two divers had been flown out that morning to make a hull inspection. But they did not get in the water. No barges or tugs hired by the owners were in sight of the ship, and none were expected at the scene that day, Hein was told.

"I don't think you're doing enough," he said to the tanker's agent. "I think you can anticipate that the Coast Guard will take over your ship."

During the night, the wind had swung around to the east and northeast, and though it was only a light breeze, the small waves it kicked up easily rolled up on the listing starboard side of the *Argo Merchant*. At daybreak the *Vigilant* was about a mile to the west of the tanker when the officer on watch spotted small globs of oil in the water floating by on the easterly winds. The leak Terry Cramer had spotted the day before had stopped later that afternoon, possibly having sealed itself as the waves ground the ship deeper into the sand. These patches were the first pollution the *Vigilant* crew had seen. They were small and scattered, and were certainly no evidence of a significant hole in the tanker; in fact, the oil may well have been sucked from the vessel by the Adapts pump. But whatever the source, oil had not appeared more than a mile from the wreck.

Ashore, Captain Hein decided he had better find out where the oil would go if much more got into the water. At 9:30 A.M. he called the Coast Guard Research and Development Center at Groton, Connecticut, and asked for a prediction of the path of a potential oil spill off Nantucket during the next several days.

About the same time, a representative of the National Response Team, which was keeping an eye on the *Argo Merchant* developments from Coast Guard headquarters in Washington, called Captain Folger to offer the assistance of the Coast Guard's Oceanographic Unit. Using special infrared techniques and a computer program, the unit could track the oil from the air, chart its daily progress, and predict where it would be the next day. Folger readily accepted the offer. If much of the tanker's cargo spilled, he thought, no service would be more valuable than to keep track of its course over the sea and predict when and where it might hit the shore.

Down in the *Argo Merchant*'s engine room, the Adapts pump hummed along smoothly throughout the early morning, making the task of emptying the hold seem as easy as sipping water through a straw. By 8 A.M., the level, which had reached about thirty feet before the pump was running, was down to less than twenty-five feet. Chambers grew more optimistic with every inch of water that went overboard.

His mood changed, however, when he learned that the other two Adapts pumps were coming out on the *Bittersweet*. He had not expected the pilots to haul any more loads out there in darkness. But at first light, he wanted to hear them chopping overhead again with another one of his units.

The air station, however, was not eager to try the job again, even in daylight. Being able to see where they were going and where they were setting the load solved only part of their visual reference difficulties and none of the logistical problems of flying and hoisting time. Flying a helicopter over the sea with an unstable load was nearly as dangerous in daylight as in darkness. In delicate hovering maneuvers, the horizon is not a good visual reference because it is too far away; a pilot needs something closer and more specific to look at. In addition, they still faced a maximum speed of fifty knots, and that still meant an hour and a half-long flight out there and a forty-five-minute flight back. Flight time alone for one helicopter getting one complete unit out to the tanker would be nearly four hours. And that didn't count loading the gear, refueling the helicopter, routine flight checks, and lowering the equipment to the deck. All told, it might take six hours to get one Adapts pump out there, the air station officials figured. The *Bittersweet* could get both of them there in seven.

But Chambers wasn't a party to that thinking. Sitting out on a tanker seventy-five miles from the air station, it was easy to lose perspective on the time it took just to get out there. Most of his oil jobs were closer to shore. Still, he didn't think the air station understood the urgency of getting the gear out to him. He got back on the radio to Cruickshank.

"I wish I'd known the Adapts were on the *Bittersweet* before she left port," he said. "I've gotta have those pumps. Tell the air station to have the pilots lift them off the buoytender now and bring them out right away."

Cruickshank got the message out, but even Chambers didn't expect anything to come of it. The superstructure of the buoytender was too high and too close to the cargo storage area on the deck for a helicopter to hover low enough to pick up one of the pallets. The issue was closed. Chambers would have to wait. He could do nothing to change it now, and he had long since learned that fretting over conditions he couldn't control on an oil spill job was a useless exercise. Besides, it was his only setback. He didn't have a barge to pump to or a tug to tow the ship off them anyway. What he did control in the operation was working just fine. By 10 A.M., the level in the engine room was down to about twenty-one feet.

Meanwhile aboard the *Sherman*, Captain Veillette was making arrangements with people ashore to evacuate about twenty-five *Argo Merchant* crewmen that had spent the night on his ship, sleeping on borrowed blankets in any open space they could find below deck. The *Sherman* wasn't equipped to put up the tanker's crew indefinitely, and Veillette wanted them ashore as soon as possible. An immigration depot was established at the Nantucket Airport, and the airlift of the crew began at noon.

Just before that, a helicopter picked up Calicchio and Chief Cramer and took them back to Boston while the rest of the damage control team, including Lieutenant Hagstrom, was ferried back to the *Sherman*. The only Coast Guard personnel remaining on board the *Argo Merchant* were the three members of the Atlantic Strike Team. Captain Papadopoulos and a handful of his officers were on board as well, but they were doing little to help the Strike Team.

By 1 P.M., the level in the engine room had reached fifteen feet. Now, however, it was holding steady. The pump kept pumping, but it was no longer making headway. Chambers was not concerned. In fact, he had pumped just about all the water he wanted out by then, anyway. If the stern lightened up too much, it could raise up slightly and drive the mid-section of the ship up harder on the shoal, he thought. With more buoyancy, it would also be more vulnerable to pounding that could open up the hull under the constant pressure of the waves. Chambers was satisfied just to maintain the fifteen-foot level until a barge arrived to hold the oil. He knew he could put another pump in the engine room and suck it dry with ease.

The crewmen of the *Argo Merchant* who remained aboard had found some cheese, bread, and sodas in the ship's store and put them out in the salon under the bridge for anyone aboard to eat. They were all right for snacks, but Chambers hadn't had anything close to a meal since the sandwich aboard the C-130 twenty-four hours before. With all the helicopters buzzing around, Chambers asked Cruickshank to see about sending over some lunch and some coffee. He thrived on coffee, but all anyone had come up with on the tanker was a Greek variety that Chambers thought was terrible. Cruickshank got to work and it wasn't long before a helicopter lowered some hamburgers and french fries and a pot of coffee to the Strike Team.

With a warm meal in his stomach and a hot cup of coffee in his hand, Chambers was happy. And he had more than a good meal to smile about. He had control of the engine room. He had more gear coming on the *Bittersweet* just over the horizon. Two tugs and a barge were on the way. The water was nearly calm, and every once in a while the sun would poke through the clouds, warming and brightening the scene for a moment as its rays glistened off the water. Chambers squinted as he stood on the deck looking out over the water and grinned. It even looked pretty, he thought. He got on the radio to Captain Veillette.

"Sir," he said, "I think we've got this one by the shorthairs."

9

The Evacuation

As the Strike Team scurried about the vessel during the night and morning, cutting off deck rails, bashing down doors, and setting up the pump, Chambers was mindful of one ticklish fact: it wasn't his ship. With just one pump in the engine room, this fact wasn't an overriding concern. He had no illusions that one Adapts would refloat the ship. The pump had saved the engine room, he thought, and it was relieving pressure on the hull. But he would need more pumps before he was likely to move the ship itself. As those pumps got closer on the *Bittersweet*, the question of control over the vessel became more significant. Chambers wanted two things: a man in charge, with authority to decide what to pump and when to pump it, and a salvage plan.

Captain Papadopoulos had left all decisions regarding salvage of the vessel up to the owners. Four owner's representatives had arrived about 3 A.M. with the second Adapts delivery, but Chambers quickly learned that none of them was empowered to make decisions either. He would have to wait for Captain Nick Skarvelis of the Amership Agency in New York.

Skarvelis was lowered to the deck of the tanker shortly before 3 P.M., along with a salvage company official. They went straight to the salon to meet with the captain, another salvage man, and the other owner's representatives already aboard.

Skarvelis took quick charge of a meeting, asked several questions, listened carefully, and within a few minutes hired the Murphy Pacific Salvage Company to run the operation. One Murphy Pacific official, Capt. Nelson Hiller, would stay aboard to take charge, as the salvage master. Skarvelis and the others with him left the ship.

Chambers, who was in the back of the room going over the ship's drawings, kept an ear on the meeting, though he did not participate. He was impressed with Skarvelis' ability to ask aggressive questions and make a quick decision. He was also happy that Murphy Pacific got the job. He knew the Coast Guard would probably hire the same company once it took over the ship and he thought the transition would be smooth. Most important, he finally had a salvage master in charge of the ship. Chambers was a Navy-trained salvage diving officer and was experienced in ship salvage, but this would be a major effort, requiring the work of a variety of marine experts. Chambers' job was setting up pumps and rigging other equipment to pump out the vessel, under the supervision of a salvage master. Like a foreman on a construction project, the salvage master hires the people and acquires the equipment needed to do the job, and he carefully formulates a plan to do it. Chambers already had some of the equipment and he and the Coast Guard had more on the way. Hiller started going over drawings, diagrams, and loading charts to prepare a salvage plan.

Within a half hour, however, the ship got word that the First Coast Guard District had invoked the Intervention Convention. The United States Government was in charge of the ship. Skarvelis' decision had quickly become moot. Hiller was left in limbo. His authority had come from the owners, and the owners no longer had a word to say about the ship. He was ready to leave the problem for the Coast Guard's salvage master and return to shore.

But Chambers wanted him to stay on. Even if a different salvage master were appointed, it could take hours for him to arrive at the scene. With the pumps on the way, somebody had to devise a plan soon to pump out the tanks. Chambers couldn't just flip a coin to decide where to put the pumps. Careless unloading of a tanker, even in port, can place intolerable pressures on its hull. The weight of the cargo in each tank, the stress it puts on the ship, and the stress resulting from removing its contents have to be carefully calculated before pumping can begin. Chambers wanted a salvage master to make those calculations.

"This international business is a problem for them on the beach," he said. "But it doesn't make a bit of difference out here right now whether the Liberians own the ship, whether the Greeks own it, whether the Coast Guard owns it, or whether I own it. We're it. We need a salvage master. You're a salvage master. We've got a pump and the *Bittersweet* ought to be here any time with two more pumps. If you work in here on the weight and balance calculations, I can work with you on the physical assessment and my guys can run the pumps, then we'd be a team. We don't need to worry about who's in charge. We've got a ship aground here with seven and a half million gallons of oil on her, and we're the only people who can do anything to get her off. I think we ought to be about doing it."

"Well, all right," Hiller said after a short pause. "Let's see what we can do."

The *Bittersweet* approached the *Argo Merchant* shortly before 4 P.M., a little later than Chambers had hoped. But the seas had started to pick up again and had slowed her down. Pete Brunk, the Strike Team officer who had supervised the moving of equipment ashore, was standing on the bridge.

"Request permission to come alongside, sir," he hollered over to Chambers.

"Yeah, come alongside," Chambers answered. "I'm not going anywhere for a while. But don't scratch the paint."

"How do you do!" McKnight added. "Glad you could make it. You guys have a nice time in town last night?"

"Yeah, y'shoulda been there," came the response from one of the other four team members aboard. "Lots of nice lookin' women. We got nice and drunk and disorderly. Just a great time." They laughed.

The Strike Team members aboard the *Bittersweet* had

been up all night, but not exploring the winter night life on Cape Cod. They had worked with Brunk, moving the thousands of pounds of equipment from the air station to Woods Hole to the buoytender.

Coming alongside the tanker was hardly like slipping up to a pier in a quiet harbor. The swells against the stricken vessel had built up to six feet. Lt. Cmdr. Jack Overath, the skipper of the *Bittersweet*, had to push her engine up to nearly two-thirds throttle to battle the swells and maneuver in a swift four-knot current that was rushing lengthwise along the hull. He powered the bow-thruster full blast to push the nose of his ship toward the tanker, and he turned the rudder so the current running bow-to-stern would push the stern toward it as well. It was the Strike Team's first look at the strength of the current along Fishing Rip and it surprised them.

The *Bittersweet* tied up to the tanker's port side at 4 P.M., and her crew immediately swung the small crane on the deck into motion. One at a time, four pallets of Adapts gear were lifted onto the *Argo Merchant*. Four of the five Strike Team members, with their personal equipment, and two scientists also stepped over to the tanker. Only Brunk stayed aboard the buoytender, to return to shore where more equipment was being flown in.

By the time the *Bittersweet* shoved off, it was 4:30, twenty minutes after sunset. It was getting dark quickly. Clouds had thickened during the past few hours, moving in on a freshening northeast wind. The seas were building. And it started to rain. The tugboat *Sheila Moran* was getting closer, within radio distance. The tugboat *Marjorie B. McAllister* with the barge was just coming out of Buzzards Bay and heading to the south of Martha's Vineyard, still several hours away.

With two more pumps aboard before the barge arrived, the salvage master knew the Strike Team couldn't use them to pump oil. But he had an immediate plan for one of the units. A senior crewman had told him the number five starboard tank was full of ballast water instead of cargo. The vessel's tank gauge plan, which listed the cargo in each of the tanks from number one starboard, center, and port through number ten starboard, center, and port, seemed to confirm the crewman's story. It showed number five starboard had no oil in it. Hiller thought the water could be pumped over the side. Emptying the tank would add buoyancy to the vessel where she was hardest aground and listing most severely.

The Adapts was the perfect tool for the job. It was specially designed and built to be lowered into cargo tanks through a twelve-inch diameter hole in the deck known as a Butterworth fitting. During cleaning operations, the hole is opened to lower a rotating spray nozzle into the tank which blasts water against the sides, washing them down. (The oily mixture then becomes part of the ship's ballast and, in one of the most controversial operations in oil transportation, is usually discharged into the ocean. More oil pollutes the seas through this cleaning method than from any other single source.)

The Strike Team, now seven strong, rigged up emergency lighting on the deck so they wouldn't have to battle the darkness Chambers, McKnight, and Darby had endured the night before. They set up a second prime mover on the port side, next to the one still running the pump in the engine room. And they erected a ten-foot-high tripod over the Butterworth fitting of the number five starboard tank from which they suspended the pump on a cable hoist similar to the one they had clamped to the ceiling of the engine room.

The number five tanks were just aft of the bridge house, along the first of six rows of tanks that stretched between the bridge and the pump room. Tanks one through four were up in the bow. With the eight degree starboard list and the flooded engine room the vessel was leaning so low astern that the deck by number nine and ten starboard tanks was under water. Waves had been washing over the deck by the number seven and eight tanks since the *Bittersweet* had arrived. By the time the pump was suspended from the tripod, the wind had kicked up the seas enough to send an occasional breaker swirling around the Butterworth fitting where the Strike Team was working, flopping the rumpled discharge hose around on the deck.

The Evacuation

With the pump ready to be lowered into the tank, McKnight opened the Butterworth fitting. Then, as two other men helped guide the pump into position, he worked a handle on the cable hoist that lowered it in short, jerky movements, as if he were letting a car down on a jack. The metal end of the pump clanked against the Butterworth fitting and stopped. Hugging the pump, two men tried to work it around in the hole, hoping it would drop through. But the pump was about a half inch too wide. It wouldn't fit.

McKnight walked over to Chambers and Hiller who were standing nearby.

"That's never happened before," he said. "It just won't go in the hole. This doesn't look like such a good idea anyway. I think there's oil in there." When McKnight had opened the Buttcrworth, he saw what looked like oil in the tank, though it was too dark to tell for sure.

"Well, the mate said it's water," Hiller remarked. "If the pump won't go in there, I guess you'll have to go in through the tank top."

McKnight looked at the tank top, a three-foot by five-foot gray cover that rose a couple of feet off the deck, then looked back at Hiller.

"I don't want to go in through the tank top," he said. We got seas crashing around up here. You don't want to open a tank top with the sea rolling in on top of it."

Chambers intervened. He didn't want an argument between the salvage officer and one of the Strike Team. He didn't think it would hurt to take a look.

"Go ahead and lift it, Mac."

The tank top opened with a ratchet-type screw which raised the cover about an eighth of an inch for every turn. The hinge and the screw itself were badly rusted, and it was a struggle for McKnight and Darby to make any headway at all.

"Watch it, Mac!" someone yelled.

He and Darby grabbed the cover and held on. A wave crashed against the ship and water rushed around them knee deep. Then, as if they were standing in the surf at the beach, the water fell off, leaving them on a clear deck. They went back to work.

After several minutes and a couple more waves, they had it raised high enough for McKnight to look in.

"It's oil," he said, gritting his teeth.

It was already bad enough sloshing around in the waves on deck, he thought. If the oil got washed out of the tank, the deck would be treacherous. The side of the ship was a downhill slide just twenty-five feet away. He thought they should close the tank top and move someplace else.

But Hiller suggested that a bulkhead might have cracked in the stress and that some oil had leaked into the tank, just covering the top of the water. He wanted the pump put in the tank.

Chambers dropped an oil gauge into the tank and lowered it on a string straight to the bottom. When he pulled it up, it showed oil. But Number Six oil gums up the gauges, he knew, and it might have registered incorrectly after getting through a layer of cargo. He went over to the rail and dropped the gauge into the ocean. When he pulled it back in, it still read oil. He had not seen any signs that the inner structure of the ship was failing, but he had to concede the possibility that there was water in the tank.

"Okay, Mac, put the pump in about fifteen feet or so," he said. "Let's see what we got."

McKnight, Darby, Jim Kuchin, Frank Williams, and Al Harker worked on the pump while Chambers adjourned to the salon with Hiller to recheck the ship's plans.

Shifting to the tank top meant that the tripod had to be moved and the discharge hoses rerigged. The cover would not open completely because of the rust, so the men had to hand feed the pump into the tank as they got slack from the hoist. The waves were hitting with more frequency and more force as the men worked. Every minute or two, they had to stop quickly, grab the tripod or a deck fitting and hang on as water crashed on the deck waist high and banged them around against the metal. Then they raced back to work to make a little progress before the next wave hit.

The Evacuation

By the time they lowered the pump into the tank, it was nearly 6:30. And for the first time during the operation, oil started surging in the tank, rising up toward the top of the tank cover as each wave passed beneath the hull. When an especially large wave hit, the oil spilled over the tank top out onto the deck.

"Okay, crank her up," Mac yelled to another team member at the prime mover. The diesel groaned into gear. The hydraulic lines filled with fluid. The pump started to whir. The six-inch hose jumped and writhed as it went firm along the deck. The fluid inside made its way to the side of the ship and poured into the ocean.

It was oil.

"Secure it," McKnight shouted.

The prime mover coughed and went off. The pump stopped. McKnight was seething. He looked down at the tank top. Oil gushed out of it onto the deck.

The rolling sea and the whipping at the end of the discharge hose had torn the manila rope that tied the hose to the deck rail. Now heavy with cargo, the hose rolled around on the deck with the oncoming waves. Al Harker wanted to tie it off again so it would not whip back and knock someone over.

As he headed out toward the edge of the ship, slightly aft of the open tank top, he slipped on the oil.

"There goes Harker!" someone shouted.

Water from another wave rushed in. Harker struggled frantically. He reached for a rope. A hose. Anything he could hold. But the wave receded and carried him toward the edge of the ship and aft, where water had inundated the deck.

Darby went after him. Just as he reached him, he fell. Both went sliding toward the rail. Jim Kuchin went after both of them. He reached them just as the wave went out from under them about ten feet from the rail. But as he helped them to their feet, another wave hit. All three went down and were carried back toward the edge of the ship.

McKnight and Frank Williams were hanging on to the tank top, helpless. They braced themselves as the wave hit them, then looked back. Kuchin, Darby, and Harker were nearly buried in the wash. McKnight thought they were gone.

When that wave passed, McKnight could see them again. They were hanging onto the hoses and braced against the rail. In a few seconds before the next wave, the three of them scrambled for the center of the deck. The ship's rail had saved them. If they had been back on the fantail, Darby thought, they would have been overboard. He had cut the deck rails from the stern.

When everyone was on safe ground, McKnight headed for the salon in a rage. He walked up to a table where Hiller was sitting, clenched his fists, and glowered at him.

"You know that tank you told us to pump that was full of water? It's full of oil."

"That's impossible," the salvage master said.

"Are you calling me a liar! I told—"

"Put it away, Mac," Chambers interjected.

"Well, there's oil all the way down. Now there's oil all over the deck. Our guys are slipping all over the place. We damn near lost Harker, we damn near lost Kuch, and we damn near lost Darby."

Chambers and Hiller went out to have a look. The oil on the deck especially troubled Chambers, and not only because of the safety of his men. When he looked into the tank top, he watched the oil surge up and down each time a wave passed beneath the ship.

He paused for a moment, glanced quickly at the other tanks nearby, then looked back at number five starboard. The surging cargo indicated the tank had a hole in it. The belly of the *Argo Merchant* was starting to open up.

"All right, screw it, Mac. Get this pump out of here and set it up back in the engine room. Let's see what we can get going back there. This thing's starting to catch up with us."

But Hiller protested. He wanted the pump dropped to the bottom. If the tank were indeed filled with oil, he knew No. 5 port was definitely empty. Instead of pumping it overboard, they could pump the oil across the deck into the empty cargo tank and still take a little of the list off the ship.

Chambers disagreed. With the tank open to the sea, no matter how small the hole, any cargo that pumped out would

The Evacuation

be replaced quickly with water. He saw no purpose in pumping against the sea.

In addition, while it sounded simple enough, shifting the gear around was not simple for the men who had to do it. The pumping system was now filled with oil. Lowering the pump to the bottom would fill up more hose and compound the problem of getting the pump out of the tank. One fifty-foot length of hose weighs nearly two hundred pounds empty. When it's full of oil, it's almost unmanageable. Chambers' men were soaking wet and cold. The temperature was forty-five degrees. The water temperature was forty-two. He just wanted to get the pump out of there, close the tank top, and work in the engine room.

But the salvage master was persistent. Chambers thought it was a point worth arguing, but didn't. Although he doubted the move would help, it wouldn't hurt either. More important, Hiller already had misgivings about his authority on the tanker. Chambers thought if he fought too hard he would lose his salvage master. The Strike Team would need his expertise when the barge arrived.

"Well, okay, we'll try it," he said.

His men started to lower the pump deep into the tank. Others, with ropes tied around their waists, ventured out to the edge of the ship to begin moving the oil-laden hose off the starboard side over to the No. 5 port tank.

Meanwhile, Chambers roamed about the ship, wondering whether the hole in No. 5 starboard was an isolated new problem or an indicator of conditions throughout the ship. He got his answer in the engine room.

At 6:30, the engine room began flooding rapidly, without warning. The Adapts pump was still running, but it was overwhelmed. The bulkhead against the pump room was flexing harder and deeper than it had been earlier that day. And the buckling along the port side passageway to the main deck was so severe now he almost had to climb over it. The Strike Team had lost control of the engine room. Chambers had stationed one man inside to keep an eye on the pump. Now he ordered him to stay on the deck outside and watch from the passageway.

"I know it's cold and shitty out here," he said. "But stay the hell out on deck. I don't want you stuck in there if that bulkhead goes."

Over the next forty-five minutes, as Chambers strode the deck in darkness, bracing himself against a cold wind-driven rain, he listened to the sound of a ship in agony. Just as the engine room had succumbed suddenly to the flooding, the tanks which had stayed intact throughout the day had abruptly surrendered to the beating. Many had opened up underwater. Now each wave smashing against the hull forced pressure into the ruptured tanks, driving the oil and air inside toward a four-inch ullage hole at the top. Many of the ullage caps had been rusted or damaged even before the Strike Team had opened them and reopened them during the day to check cargo levels. They couldn't withstand the compression within. The air rushed through the holes with a deep, eerie whoosh. Then, as the trough of a wave passed beneath, creating a vacuum inside the tank, air was sucked back in through the opening. Caught in the rhythmic, relentless flow of the ocean swells, the *Argo Merchant* was howling.

The oil also was hurled against the top of the tanks, hitting with a muffled smack, then squirting through the ullage holes. The goo shot into the air and then oozed over the deck, as if giant tubes of black toothpaste were stored in the hull and had suddenly been squeezed.

One stubborn ullage cap that had been pinned shut with two three-quarter inch bolts, blew right off. Oil shot out of that hole like a geyser, reaching fifteen feet in the air. Chambers watched the display as much with fascination as with sadness in losing so rapidly what had taken a whole day to accomplish.

While he was watching the tanks, he got word that the *Sheila Moran* had arrived on the scene. It was the the tug without the barge, and her skipper wanted to put a wire on the tanker and try to pull her off the shoal. But Chambers said no. They had lost control of the ship. If the tug had arrived two or three hours earlier, Chambers thought, he would have done it. They could have attached the cable and kept a constant strain on the ship while his team put a second Adapts in the engine room and the third in number five starboard. He would have asked permission to pump oil over the side, and then pumped and pumped until the tanker came free. He had figured they would have to get rid of a half million gallons overboard to lighten her enough to get her off, and he wondered with a smile how his request would have been greeted on the beach.

But all that was idle speculation now. Just a few hundred feet off the shoal, the water was 120 feet deep. If they pulled her off into water that deep, in her flooding condition, Chambers thought, she would sink in an hour.

Back at number five starboard, the other members of the Strike Team were unable to work. Ten-foot breakers were rolling in on the deck, where the tripod was set up, and crashing on the catwalk in the center of the ship, about forty feet from the rail. The men had managed to get the discharge hose into the number five port tank, but the surf and the venting made any effort to continue futile. They couldn't even reach the pump to get it out of the tank. Chambers told them to forget it.

"Start tying some of this other gear down so we don't lose it," he said. "We're not getting any more work done tonight."

While Chambers was touring the ship, Captain Hiller made the rounds as well, on his own. By 8:30 P.M., he had seen enough. The venting tanks, the twisting and flooding engine room, and the increasingly poor weather had convinced him it was time to get off the tanker.

Chambers had not quite reached the same conclusion. He and his men were equipped to spend the night. They could alternate a watch to monitor the condition of the ship and catch some sleep in the salon, he thought. They had rubber rafts aboard if they had to get off in a hurry. And the *Vigilant* was close by if they needed to be rescued.

But he had nine civilians on board, including the captain and five other crewmen, two scientists, and the salvage master. They certainly were not mentally geared to stay the night on a howling, rocking ship, he thought. He also had to consider the helicopter pilots. They would have to come way off shore again in the middle of the night. The weather was getting worse. If he waited too long, he thought, the weather might shut down the air station. Finally, there was the nagging problem of the engine room. With the twisting and flooding back there now, he would not be surprised if it broke off during the night. That could make the rest of the ship so unstable in the current swirling around it, Chambers thought, that it could roll over.

Shortly after 9 P.M., he called the *Vigilant* and the *Sherman* to talk to Cruickshank and Veillette.

"Everything's gone downhill over here," he said. His voice was calm and matter-of-fact. "We lost the engine room a couple hours ago, lost six feet in an hour. The tanks are venting and pushing oil through the ullage caps almost fifteen feet in the air. The ship's starting to twist slightly, and with the increased weight in the engine room, she could break off in these seas. Also, we got a pumproom bulkhead with oil behind it flexing down in the engine room. If that goes, we'll have one incredible mess. I thought we had it, but the damage in the past few hours since the weather picked up has been extensive. I think we better get everybody off here while we still can."

"Okay, Barry," Cruickshank said. "We'll get a couple of helos over there right away." He figured that if Chambers wanted to get off, it must be in pretty bad shape.

Cruickshank maneuvered the *Vigilant* in as close as he could to the tanker. He knew he would land some of the passengers and he wanted the helicopter making quick, short trips back and forth to get the men off. Meanwhile, on the *Sherman*, Lt. John Painter prepared to launch the small H-52 off the deck while the cutter radioed the air station to ask for an H-3 as well.

Chambers told his team members and the other people aboard to gather their personal gear and prepare to evacuate. And he ordered everyone back aft, near the buckled passageway into the engine room, where they would have some shelter from the wind and rain until they were lifted off the deck.

Painter was hovering over the tanker about 9:45. With the Strike Team's emergency lights burning, the pilots had a much better visual reference to the deck than the night before. He lowered his rescue basket to the deck.

Chambers tapped McKnight, Darby, and Charlie White, a third Strike Team member, for the first ride off.

A twenty-five-knot wind was already blowing water and bits of oil through the air, and now the wash from the whirring helicopter blades stirred it up even more. Gobs of oil slapped the men in the face and back as they scrambled one by one into the rescue basket for the hoist to the helicopter. At 9:56, Painter headed for the *Vigilant* about five hundred yards away. The H-3 was already underway from the air station.

As Painter hovered over the *Vigilant* a few minutes later, an icy film on his windshield smeared his vision. The cutter flight deck below him was pitching and rolling violently. He would have to be quick. If the deck was slanting too much, he risked sliding off into the water.

Painter hesitated for an instant as he got near the deck and the yellow-lined landing grid in its center. He switched on his landing lights. The chopper dipped up and down, as if Painter were trying to feel his way down. It jerked back and forth as it was buffeted by winds. Cruickshank held his breath. He had never seen a pilot put landing lights on over the deck. The *Vigilant's* spotlights plus a landing officer with flashlights should be plenty of illumination, he thought.

The helicopter touched the deck, bouncing lightly, then setting down firm. It was out of the grid. One wheel was at the edge of the deck.

"Make it!" a crewman yelled to the passengers.

McKnight, Darby, and White unbuckled their seat belts, grabbed their gear, and bolted out the cargo door. Painter reached out and ran his finger across the windshield. It wasn't ice. It was oil.

As the last man was out, Painter gunned the engine and the helicopter rose quickly and veered off toward the tanker. Cruickshank was relieved. It was the toughest landing he had ever seen on his flight deck.

Painter went back to the tanker and picked up four more crewmen. This time he decided to head for the *Sherman,* with its larger, more stable deck. Through his rain-splattered, oil-smeared windshield, he looked down on a vast, black ocean interrupted only by a few blurred white patches marked by the running lights of the vessels below. He settled into a hover over one patch of lights and descended slowly toward the deck. An urgent voice broke the silence on the *Vigilant*'s radio.

"Christ, a helicopter is trying to land on me!"

It was the skipper of the *Sheila Moran*. Painter had picked the wrong blur. Cruickshank quickly called the pilot on the radio to tell him it was not the *Sherman*.

Painter was not concerned. He had just gone down for a closer look at what was below him. But as he headed off toward the large Coast Guard cutter, the skipper of the tugboat sighed heavily and wiped his brow. To him, at least, it was a close call.

By the time Painter set down on the *Sherman*, the H-3 from the air station was on the scene. The larger helicopter could get the remaining nine passengers in one trip, so Painter shut down for the night.

For the H-3, the mission was more routine. Chambers continued to direct the evacuation on the tanker, calling out passengers from the passageway each time the basket hit the deck. Finally, just he and Captain Papadopoulos were left. The master of the stricken vessel had hesitated. In fact, Chambers had had to send one of his men back to the salon to get him, not once, but twice. The captain still didn't want a tugboat attaching a line to his ship. He had gathered three suitcases full of his belongings. Despite the captain's wish to be the last one off, Chambers didn't think Papadopoulos could have handled the basket and all his luggage alone. He helped the captain in, tossed in the suitcases, and sent him up.

About 10:30 P.M., as the basket came down for the final time, Chambers was standing in water halfway up his calves. His face and hair and hands were covered with oil. He threw his gear into the basket, climbed in, and ascended, twisting into the darkness.

As he looked down, the emergency lights, still burning brightly, cast an eerie pall over the now empty deck. No one was aboard to hear the whistles and the howls from the cargo tanks or to feel the vibrations of the waves pounding against the hull. The *Argo Merchant* was abandoned.

Five and one half hours later, at 4 A.M., the *Marjorie B. McAllister*, with a barge in tow, arrived at the scene of the grounding. The weather had gotten there first.

10
The Strike Team

As the men headed back to the air station, the stench of oil hung thick in the helicopter, more noticeable to the pilots than to those who had lived with it aboard the tanker for the past day and a half. Oil dripped off the boots and clothes of the passengers in the cabin. It clung to the webbing of the seats, stained the seat belts, and covered the floor, as if someone had knocked over a barrel of roofing tar.

At 11 P.M., the helicopter landed at the air station. The pilot shut down the engines and a silence fell over the cabin for a moment. But before the men could unbuckle themselves from their seats, the stillness was interrupted by a meowing sound coming from a burlap bag near Captain Papadopoulos. With all the papers, personal belongings, and ship's documents he had crammed into three suitcases, the master had not forgotten the ship's cat. The rest of the passengers smiled skeptically, shrugged, and made their way out of the cabin.

Chambers was also covered with oil. His face was spattered and blotched, as if he suffered from a rare disease. His hair was matted and his hands, arms, and coveralls were black. After checking in by telephone with Captain Hein in Boston, Chambers searched around the air station buildings for a place to clean up.

The Strike Team uses two chemicals to remove the oil from their gear and their bodies. The equipment is easy. They spray it with a strong solvent, let it stand for a few minutes, then wash it down with a garden hose. The pressure of the water knocks pieces of oil away.

Cleaning their skin, however, is a much longer, almost ritualistic process. The team sets up a cleaning area complete with rags and pails and puts down a series of absorbent pads which are made for soaking oil off water or beaches. They step on one of the pads, strip, and step over to another pad where they take rags and rub themselves down with an emulsifier not quite as strong as the solvent they use on their equipment. They wipe themselves off with a clean rag and wash themselves once again with the emulsifier. Then they take a shower. It usually takes them an hour and a half from start to finish.

Chambers had three layers of clothes on as he worked aboard the *Argo Merchant,* his coveralls over a cold-weather suit over thermal long underwear. The coveralls were caked with oil, and the insulated suit was saturated. Even the thermal underwear was stained brown. In warmer weather, when he doesn't wear as much clothing, it is his skin instead of the underwear that turns brown. This time, however, he had only to rub down his head, hands, and arms with emulsifier before getting into a soothing hot shower. Sometimes the men wear disposable rain gear on the job. When they clean up, they use insecticide sprayers full of the strong solvent to wash each other down, hit each other with a garden hose, then step out of the rain slickers. It keeps some of the oil from penetrating their clothing.

After he joined the Strike Team, Chambers, who was bald at twenty-seven, paid a specialist $2,000 to implant hair in his scalp, suturing each strand in place. But the hair wouldn't hold up to the frequent dunkings in oil and the baths in solvents. Chambers didn't like it anyway, and within a year, he had it all removed. The specialist did that for free.

After two or three days at a spill, the nightly cleaning treatments leave their skin so dry it turns white and flakes. So after the onslaught of oil, solvents, soap, and water, the men treat their skin with moisturizing cream.

Less abrasive cleaning products are on the market, but they are designed for auto mechanics who want to wash their hands after a day of tinkering with engines. They just aren't strong enough for the Strike Team. The men have to rub their skin raw for two hours before the weaker chemicals work, and most of them have decided they would rather use the stronger ones, get cleaned up quickly, and get out to have a beer.

A couple of men have had allergic skin reactions to the solvents, and Chambers himself had trouble breathing for several days at one spill because he couldn't get the smell of the chemicals out of his lungs.

The men also have had difficulty washing them out of their skin. At a spill in Cleveland, they cleaned up, took a shower, and went into a swimming pool at their hotel. Some of the men sat for a while in a whirlpool bath around the pool and the warm water sweated the chemical out of their bodies. The room was permeated with a pungent chemical odor and other hotel guests wryly accused the Strike Team of leaving an oil slick in the swimming pool. At another spill in Cleveland, Chambers once returned to his hotel room to find the maid had forgotten to make his bed. His pillow case and sheets were smeared with a light brown residue. On at least two occasions, hotel managers, fed up with oil-slicked bedding and the stench of cleaning solvent, have asked the Strike Team to leave.

Like the rest of the nation's oil spill fighting apparatus, the Strike Team grew out of the *Torrey Canyon* accident, in which that Liberian tanker, carrying thirty-three million gallons of crude oil—nearly four and one half times as much as the *Argo Merchant*—ran aground and broke up off the coast of England in March, 1967. All of her cargo spewed into the sea, and the futile efforts of the English and French to cleanse the fouled beaches and treat oil-soaked birds captured the attention of the world. There had been many oil spills before then and there have been thousands since, but the *Torrey Canyon* was the first international oil disaster and it helped give birth to the cause of the environment. It still stands as the largest oil spill on record.*

Kenneth Biglane of the old Federal Water Pollution Control Administration, a forerunner to the Environmental Protection Agency, was one of a team of U. S. officials who went over to England and France to observe the cleanup operations and to consider what the United States would do if a similar incident occurred off her coast. When the observers, who also included Coast Guard officials, returned home, they began drafting a contingency plan by which a variety of federal agencies might combat an oil disaster such as the *Torrey Canyon*.

Meanwhile, within two and a half months of that incident, President Lyndon B. Johnson directed the Secretary of Interior and the Secretary of Transportation to assess the nation's ability to protect the coastline from oil pollution and to recommend a national and international program aimed at preventing spills and effectively cleaning them up when they occur. Referring to the *Torrey Canyon,* the President said it is "imperative that we take prompt action to prevent similar catastrophes in the future and to insure that the nation is fully equipped to minimize the threat from such accidents to health, safety, and our natural resources."

The secretaries' report was issued within a year and concluded: "This country is not fully prepared to deal effectively with spills of oil or other hazardous materials—large or small—and much less with a *Torrey Canyon* type disaster. Because sizable spills are not uncommon and major spills are an ever-present danger, effective steps must be taken to reduce this nation's vulnerability."

Indeed, the nation's oil pollution fighting capability was in disarray. Of three laws that applied to the problem, only one

---

*On March 16, 1978, that record was surpassed when the 220,000-ton *Amoco Cadiz*, an American-owned, Liberian-registered supertanker, suffered a steering breakdown in a storm and ran aground off Brest, France. The tanker broke in half and eventually spilled her entire cargo of 68 million gallons of crude oil, killing waterfowl and marine life and despoiling at least 100 miles of French beaches, some of which had been fouled by the *Torrey Canyon*'s cargo eleven years earlier.

dealt with cleaning up spilled oil: the Oil Pollution Act of 1924. But it was virtually unenforceable because it made only "grossly negligent or willful" discharges of oil illegal and such motives are difficult to prove. It also specifically exempted spills from shore-based facilities as well as those resulting from unavoidable shipping accidents, collisions, and strandings. And it provided no money for a federal agency to spend to clean up oil spills. In fact, forty-four years after its passage, regulations for enforcement of the law still had not been devised. Coast Guard officials felt the law was useless.

While the Coast Guard, the Water Pollution Control Administration, the Army Corps of Engineers, and other agencies all had the potential to devise plans to battle oil spills, none of the potential was coordinated. It was rigidly stratified in a bureaucracy that was more accustomed to creating another agency to solve a problem than to pooling the resources already available in existing agencies. Extensive inventories of cleanup equipment were available in some areas, particularly along the Gulf Coast where the oil industry was well entrenched. But vast stretches of the coastline and inland lakes and rivers had almost no immediate access to containment booms and skimmers. None of what was available, even along the Gulf Coast, was effective anywhere but in quiet harbor waters.

No federal agency had developed any techniques or equipment to respond quickly and effectively to a disabled tanker such as the *Torrey Canyon* and pump her oil into another vessel before it spilled into the sea.

One of the recommendations crammed into the sixty-one-page report to the President was that legislation be enacted giving the federal government authority to initiate cleanup measures at any oil spill no matter what its source and to recover the costs of such an effort from those responsible for the spill. The report suggested a revolving fund of twenty million dollars would be needed to ensure that this responsibility could be carried out rapidly enough to prevent a potential spill or to begin cleanup operations soon enough to alleviate at least some of the devastating environmental effects of the pollution. And it urged the development of new equipment to contain and recover spilled oil as well as a means to get the equipment to the scene of a spill quickly.

The report made no specific mention of a special team to respond to oil spills, but it noted that contingency planning on the local, regional, and national levels was moving forward at an urgent pace. Among the people doing that planning were Biglane of the Federal Water Pollution Control Administration, and Cmdr. Dan Charter, who headed the Coast Guard's fledgling environmental protection program. Charter was thinking seriously of including within a national plan a Coast Guard oil disaster team, trained and equipped to fight major oil spills.

The result of their work was an interagency agreement announced by President Johnson in 1968 shortly before he left office. It listed how each of a variety of federal agencies would be mobilized in the event of a major oil spill. The key to the agreement was a division of the overall authority for fighting spills between the Coast Guard and the Water Pollution Control Administration. The Coast Guard would be in charge of any spills along the coastline or in the Great Lakes. Biglane's agency would take command at inland spills. An oil disaster team would be established to assist clean up operations at any spill—inland or along the coast.

But just working out the agreement and getting it on paper didn't mean the United States suddenly had an oil spill team ready to jump aboard the next *Torrey Canyon,* or equipment to contain oil spilling from a barge in the Hudson River. Regional and local plans had to be refined and money had to be spent. The planning process within the Coast Guard stalled late in 1968 when the Johnson Administration shifted planning responsibilities to the Department of Interior. The interagency agreement still had no firm basis in law and Dan Charter and other Coast Guard officials felt they were on shaky ground asking for money to set up an oil spill team. To resolve both problems, the elements of the interagency agreement were submitted to Congress as proposed amendments to the Federal Water Pollution Control Act.

On January 28, 1969, a Union Oil Company offshore oil production platform suffered a blowout off Santa Barbara, California, spreading the first of more than 700,000 gallons of oil into the seas toward the exclusive Santa Barbara beaches six miles away. Lt. George Brown, commanding officer of the Coast Guard Group office at Santa Barbara, was the federal official in charge of cleanup operations as the oil reached shore. But he could do little more than supervise pollution companies hired by Union Oil who jammed the beaches with pay loaders, bulldozers, and oil drums for several weeks to clean up the mess. No federal law gave Brown or any other Coast Guardsman the authority to get his hands dirty and help clean up at an oil spill. Without a law, he had no money to do anything, despite enormous public pressure to act. In fact, members of his staff questioned whether they were even authorized to install additional telephones to handle the barrage of inquiries about the spill. Brown had to rely almost exclusively on Union Oil's attitude and approach to the clean up and on his own powers of bluff when he wasn't happy with the company's effort.

He wrote an extensive report about the spill based on his experience there and on discussions with other federal officials interested in oil pollution, including Dan Charter. Among the report's recommendations was one to establish a Coast Guard oil-pollution fighting team.

The public outcry resulting from the Santa Barbara incident helped spur Congress to action the following year when it considered the amendments to the Water Pollution Control Act. But it was a lesser known incident, the grouding of the *Delian Apollon,* which helped Dan Charter win his battle for the oil spill team.

On February 13, 1970, a harbor pilot boarded the tanker at the entrance to Tampa Bay, Florida, to guide the vessel to the Florida Power Corporation's Weeden Island plant near St. Petersburg. Thick fog had reduced visibility to one quarter mile on land. But the air over the bay was clear and the pilot led the ship and its cargo of four and a half million gallons of fuel oil toward the mouth of a narrow channel leading to the plant. Within two minutes of entering the channel, the ship was

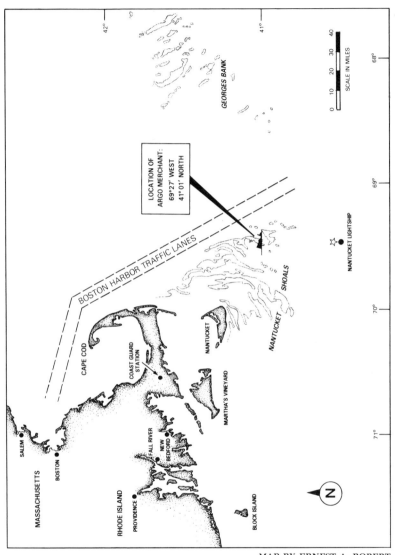

MAP BY ERNEST A. ROBERT

1. A few hours after the tanker's grounding on December 15, 1976, members of the *Argo Merchant*'s crew stand on the stern, looking up at Coast Guard helicopters which have come to their aid. (U.S. COAST GUARD)

2. December 16, 1976, the day after the grounding, a Coast Guard helicopter hovers over the bow. Coast Guard workers are on the bow below. (U.S. COAST GUARD)

3. December 16, 1976, the Coast Guard cutter *Vigilant* keeps a close watch on the *Argo Merchant*. (U.S. COAST GUARD)

4. Saturday, December 18. Waves breaking over the main deck of the *Argo Merchant* make it impossible for members of the Coast Guard's Atlantic Strike Team to get aboard to continue salvage efforts. (U.S. COAST GUARD)

5. Sunday, December 19. Lt. Cmdr. Barry Chambers, commanding officer of the Atlantic Strike Team (left) and one of his men, covered with oil, are standing in front of the *Argo Merchant*'s anchor chain. They are trying to get the anchors out to stabilize the bow—prevent it from jerking back and forth in the surf. (U.S. COAST GUARD)

6. December 21, 1976. After sustaining punishment from the waves for six days, the *Argo Merchant* broke in half, spilling most of her cargo—7.6 million gallons of heavy industrial heating oil—into the ocean. (LAWRENCE S. MILLARD, *Providence Journal*)

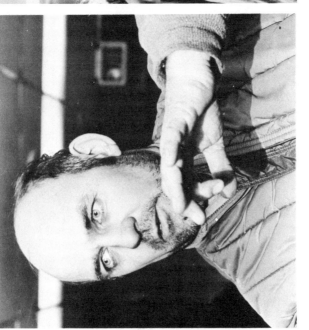

7. December 22, 1976. Chief Warrant Officer Peter Brunk of the Atlantic Strike Team at the Coast Guard Air Station on Cape Cod talks about the effort to save the tanker and her cargo. (ROBERT EMERSON, *Providence Journal*)

8. December 22, 1976. Barry Chambers gets ready to leave the Coast Guard Air Station, Cape Cod to attend a U.S. Senate hearing in Boston called by Senator Edward M. Kennedy of Massachusetts to look into the oil spill. (ROBERT EMERSON, *Providence Journal*)

9. December 28, 1976. After breaking for a second time on December 22, the bow of the *Argo Merchant* flipped over and slowly drifted away from the stern section. (U.S. COAST GUARD)

10. Clement Griscom of the University of Rhode Island's Coastal Resources Center, one of dozens of scientists who tracked the spill, indicates the slick's position on a nautical chart. At his feet are special markers which were dropped into the water near the oil slick to help scientists follow its path and predict where it might come ashore. (WILLIAM K. DABY, *Providence Journal*)

11. A special mixture of flammable chemicals dropped onto a large oil pancake ignites but fails to consume much of the oil spilled from the *Argo Merchant*. Several attempts to burn the spilled cargo were unsuccessful. (U.S. COAST GUARD)

12. December 31, 1976. Two weeks after the grounding, just the tip of the hull (foreground) and a piece of the stern section remain above the water. Within another week, all but the kingpost (center, right) had disappeared beneath the surface. (U.S. COAST GUARD)

shrouded in fog. Two minutes later she was aground. Two of her cargo tanks were pierced and twenty-thousand gallons of Number 6 oil leaked into the water.

Florida officials were angry over what they considered an inept Coast Guard response to the spill, and one man who felt the pressure was Rep. William C. Cramer, a conservative Republican Congressman from the Tampa area. He held hearings to find out why the government had done so poorly and to see what could be done to improve the performance the next time. Charter decided the time was right to push to get the elements of the interagency agreement anchored in federal law.

Cramer agreed that an oil spill team was necessary. But with the popularity of the Green Berets, who were performing daring missions in Vietnam during the late 1960s, Cramer was enthralled with a commando-style operation, and that was what he wanted reporting to oil spills.

"I want it called a Strike Force," he told Charter.

"Well, perhaps you could write that into the law and make it one of the requirements of a national contingency plan."

Cramer agreed. The 1970 amendments to the Federal Water Pollution Control Act were passed and required the President to publish a national contingency plan for oil spills. The plan, the law said, shall include the development of a National Strike Force capable of reporting quickly to major oil spills to contain them and clean them up. The interagency agreement, with minor revisions, became the plan. And Charter had what he needed to begin budgeting in earnest for a Strike Force. But even without that absolute authority, he had already dabbled with the idea.

Howard Wirt was an executive petty officer stationed aboard a tugboat in Norfolk, Virginia, in 1969 when he became interested in pumping equipment that some Coast Guardsmen and civilians were testing in the water near the tug. He got talking with the officer in charge, and before long he had a set of temporary duty orders to work full time on the project. The pumping equipment was the prototype of the Adapts system; thus Wirt became, in effect, the first member of the Strike

Team. On the assignment he reported to both the man in charge of the research project and to Charter, whose environmental post at Coast Guard headquarters in Washington was within the Office of Intelligence and Law Enforcement. It didn't sound like it had much to do with pollution, but the unit's Coast Guard acronym was OIL.

By 1971, Wirt was reporting to oil spills regularly, as much to work out problems with the pumping as to clean up oil. He had also picked up a couple of other men along the way, both of whom were officially stationed with him in Norfolk. In a move that was a slight bending of the rules, the men arranged a "temporary" transfer to work with Wirt by unobtrusively typing up orders on a teletype message and by explaining to their superiors that Coast Guard officials in Washington wanted all three of them working on the pollution project. No one ever questioned them and, unofficially at least, the Coast Guard had not just an oil spill man; they had a team.

Meanwhile, Charter tinkered with federal pollution statutes and Coast Guard budget regulations and won an appropriation for six official oil team positions, with a headquarters in New York. Some of those positions were filled, but it was more than a year before Wirt and the others got orders to report to New York. They continued to respond to spills from their temporary assignment in Norfolk, working out of the nearby Coast Guard Air Station in Elizabeth City, North Carolina.

During his first two years with the team, Wirt averaged three hundred days a year on the road. The unit was covering the whole country, and he went from one spill to the next, often without even coming home first. He was in Portland, Maine, in 1972, where the Norwegian tanker *Tamano* struck Soldier's Ledge at the entrance to Casco Bay, ripping open its hull and spilling 100,000 gallons of oil. He went to southern Utah in 1972, where a broken sixteen-inch pipeline spilled more than 285,000 gallons of oil into the San Juan River, threatening the waters of an Indian reservation and a national recreation area. The team fought heavy rains, floods, and snowstorms in their effort to clean the oil from the river. He went to Pottstown, Pennsylvania, the same year after torrential rains from Hurricane Agnes forced the overflow of eight million gallons of waste oil stored in lagoons along the Schuylkill River. The team found oily sludge twenty feet high in trees and on the second floor of some buildings.

In 1973, he was on the southwest coast of Puerto Rico, where the *Zoe Colocotroni*, a Liberian vessel, ran aground with seven and a half million gallons of crude oil. Instead of waiting to pump the oil into a barge, her captain immediately pumped two million gallons of cargo overboard and floated her free. The blanket of oil spread over beaches and into mangroves; island residents waded into the water to help the Strike Team maneuver floating booms into place to collect the oil where it could be sucked up by vacuum trucks. One day while Wirt was having trouble getting his directions across to a group who spoke only Spanish, another member of the team claimed enough proficiency in the language to help him out. But when he hollered to the men, they let the boom go, walked out of the water, and sat down on the beach. He had mistakenly told them to go take a break. And Wirt was in Cold Bay, Alaska, where a tanker transporting aviation fuel to a refueling depot ripped a ninety-three-foot gash in her hull when she hit a reef. The Adapts system was designed to pump oil, but Wirt didn't know whether it would safely pump a volatile cargo like aviation fuel. Only two men at a time were allowed on the deck of the stranded vessel while the equipment was set up. When he turned on the pump, Wirt put his fingers in his ears, bracing himself for an explosion. Instead the Adapts worked.

Meanwhile, it had become clear that the need for personnel to battle the spills had been seriously underestimated. Using the authority gained in the 1970 pollution law amendments, Charter had requested three teams of eighteen people to cover the Pacific, Gulf, and Atlantic Coasts. By 1973, the three separate teams were established. The Atlantic Strike Team was shifted to Elizabeth City. The Gulf team set up in Bay St. Louis, Mississippi, and the Pacific unit was established in San Francisco.

Barry Chambers was in New York in 1971, a lieutenant in the Coast Guard, preparing for a two-year assignment to electronics school when he first learned of the Strike Team. The idea was still mostly on paper. One of the men assigned to the New York headquarters asked him about setting up a diving program within the Strike Force. Chambers agreed to draft a proposal. He also decided he was interested in joining the team. But a friend who was a Coast Guard captain suggested he wait. They'll be fussing around for a couple of years trying to figure out what they're doing, he told Chambers. Go to electronics school while they get it off the ground. Then join.

It was good advice. He went to school four hours a day, wore civilian clothes, got paid, and had no other duties. So, to keep himself occupied, he got a part-time job with a commercial diving company, worked on small diving projects for the Coast Guard and, eventually, went to oil spills when the Strike Team was called to action. The commercial diving work earned him extra money. Both spare-time diving jobs gained him extra experience to go with his salvage diving training and some extra equipment to build up an inventory of diving gear. The trips with the Strike Team fostered an interest in cleanup technology and gained him an insight into how the team worked. The extra Coast Gaurd duties were a primary reason he earned a commendation medal during that time. He also was named Junior Officer of the Year. His leisure time activities sometimes took precedence over his official Coast Guard assignments, and once in a while he had to cut a few electronics classes to get to an oil spill.

The extra work was almost second nature to Chambers. And it was fun. Shortly after he joined the coast Gaurd in the late 1950s, he almost worked himself to death having fun. A hurricane had hit the Outer Banks of North Carolina, where the Coast Guard had hundreds of miles of telephone lines for its own communications system. As an electronics technician, he was one of two men who led work crews repairing the lines damaged by the storm. For several weeks, he worked from 5 A.M. to 10 P.M., climbing poles, restringing wire, and living on a piece of toast for breakfast, peanut butter crackers and a Pepsi for lunch, and a hamburger for supper. One day he passed out on a ferry boat taking him between islands. He went back to the Coast Guard station and crawled into bed, unable to talk. A local doctor told him he had laryngitis. Three days later, he was so dehydrated he was rushed to a hospital in Norfolk, where he spent five days in an oxygen tent. Doctors told his family he was close to death. No warrant officer or lieutenant had pushed Chambers that hard. He just liked climbing telephone poles. And at eighteen, he didn't understand the balance between working, eating, and sleeping.

By the time he got out of electronics school in 1973, the Atlantic Strike Team was looking for an executive officer, the second in command, who knew something about salvage. As a trained salvage diving officer who had learned a lot about the Strike Team, Chambers seemed not only the logical but the perfect choice for the post. The electronics people at Coast Guard headquarters, however, didn't think so. Chambers had been in that field throughout his career, and the Coast Guard had just sent him to school to learn even more. They weren't about to lose him to a pollution team. But Chambers looked back on fifteen years in the Coast Guard in which he had been a diver, had served uniquely as both a deck officer and an engineer aboard a polar icebreaker, had built loran stations, and had helped replace many of the Coast Guard lightships with mammoth permanent navigation buoys. He had never had a job in the Coast Guard he didn't want. He didn't want an electronics job in 1973. He wanted the Strike Team job. He got it.

For the next three years, Chambers worked with Hugh ("Skip") Williams, the commanding officer of the team, to build the unit from its status as a telephone number and a couple of Adapts pumps into a full-fledged, battle-trained outfit. Williams, Chambers thought, was an excellent administrator who enjoyed desk work and who was politically astute. He knew how to reach the right people with the right messages in soliciting support for the Strike Team. Chambers himself did not like sitting at a desk. He liked working with the equipment. He spent a lot of time scrounging for gear, particularly surplus equipment from other branches of the military. Boats, barges, trucks, cranes, and outboard motors started piling up around the Strike Team's headquarters. He liked the excitement of working at a spill and was the diving and salvage expert on the team. He and Williams were a perfect balance, he thought. And for the first year, they also had Wirt, who was the oil spill cleanup expert. He imparted his skills to Chambers and other members of the team before he was transferred in 1974.

The Strike Team also had someone helping them in Washington. George Brown, who had been in charge of the Santa Barbara spill, was transferred to Washington late in 1971 to take Dan Charter's job. After having worked at the nation's most publicized spill with no Strike Team and no money, he knew the value and importance of both. But there were others in the Coast Guard who didn't. The notion of an elite team of troubleshooters storming like commandos into a crisis went down hard among some district commanders and captains of ports—those responsible under regional contingency plans for cleaning up oil spills. They were proud men who felt that if they had to seek help from another unit to carry out their duties, they had failed. Many were reluctant to call the Strike Team. A few openly tried to kill it. Brown battled to get the three Strike Teams at the scene of every significant pollution incident in their areas. At first, it was almost as important for the Strike Team and the local officials to get to know each other, he felt, as it was for the oil to get cleaned up. Once a local captain of the port was convinced of the value of the team, Brown used the officer's testimony to convince others. Slowly, he wore down the opposition with success stories. But most important, in Chambers' view, Brown won the support of two admirals at headquarters and outmuscled the opposition politically. That cleared the way for budget appropriations that were essential to the survival of the team and to cementing its place in the Coast Guard.

On the night of August 9, 1974, the 206,000-ton Shell supertanker *Metula*, carrying 194,000 tons of Arabian crude oil, ran aground on a shoal known as Satellite Patch in the Strait of Magellan off the coast of Chile and tore open 260 feet of her

1,068-foot hull. Although her position was especially precarious, and her total cargo, if spilled, would easily surpass the magnitude of the *Torrey Canyon*, the grounding received little notoriety. That was partly because she was in a barren, sparsely populated section of the world not easily accessible or sufficiently well known to draw much attention, and partly because most of the world was preoccupied with another momentous event that occurred the same day: President Richard M. Nixon resigned.

Within two weeks, while the mammoth tanker lay on the shoal leaking her cargo, the U. S. Coast Guard was asked to assist those already trying to salvage her. On August 27, an eight-man contingent representing all three Strike Teams arrived with three Adapts pumps. They worked for fifty-seven days under conditions so primitive they went weeks without showers. During one three-day period, winds blew ninety to one hundred miles an hour. The Strike Force played a critical role in the effort, which led a month later to the Metula's floating free and to the eventual pumping off of 140,000 tons of her cargo. The other 54,000 tons spread over the strait, killing wildlife and coating seventy-five miles of wilderness shoreline with a strip of oil fifty to two hundred feet wide and one to four inches thick. The spill itself was never cleaned up.

Apart from cooperative efforts with Canada on spills along the U. S.–Canadian border, the *Metula* was the National Strike Force's first international job. The second came five months later, when the *Showa Maru*, another supertanker, which flew the Japanese flag, went aground in the Malacca Strait near Singapore laden with 237,000 tons of crude oil. Four members of the Atlantic Strike Team were among the U. S. contingent called to help with that spill. The vessel was refloated and towed to safety after spilling 4,500 tons of cargo.

But there was plenty of work at home as well. In fact, while four men were on the *Showa Maru* from January 14 to February 15, 1975, other members of the Atlantic Strike Team were at spills in Marietta, Ohio, where gasoline had leaked from storage tanks into the ground; in St. Croix, Virgin Islands, where the barge *Michael C. Lemos* discharged 136,000 gallons of crude oil into Limetree Bay; in Philadelphia, Pennsylvania, where the oil tanker *Corinthos* exploded and burned after colliding with another tanker, the *Edgar M. Queeny,* killing three men and spilling as much as thirteen million gallons of oil; and in Portsmouth, New Hampshire, where the *Athenian Star,* battered during a rough trip to Rotterdam, leaked about one third of its 5.5 million-gallon cargo from a crack leading from a fifteen-foot hole in her bow.

In between spills, the Strike Team is constantly working with their gear, either on training drills to increase their efficiency in loading it onto airplanes and setting it up at spill sites, or in testing it, repairing it, and rebuilding it, to make sure it is ready when the next call comes. Many of the members are volunteers, like Lt. John Clay, who got a degree in oceanography and joined the team because of his interest in the environment. Chambers hand-picked others, like McKnight, who is a diver and an expert mechanic. The first thing Chambers tells a prospective member of the team is to forget about 8 A.M. and 4:30 P.M. Some shore commands in the Coast Guard work those hours, but the Strike Team knows no such convenience. It is on standby twenty-four hours a day and at spills they sometimes work three days straight without sleep. The second thing he tells him is to forget about the traditional division of responsibility in the military, where a damage controlman welds and a boatswain ties knots and neither thinks of doing the other man's job. Part of the concept of the unit is teamwork, and to function efficiently, Chambers believes, each member has to know at least a little about the other members' jobs. If McKnight, who runs the Adapts pumps, is at a spill on the St. Lawrence River and a tanker goes aground in Chesapeake Bay, someone else might have to run the Adapts.

But while he makes those demands, Chambers tempers them with a flexibility not always built into a military system. There is no time clock at the Strike Team's headquarters. If four of his men have been working out of town on a spill for three weeks straight and return to Elizabeth City on a Saturday afternoon, he doesn't expect them in at 8 A.M. on Monday. Each man is issued $500 every three months to pay for hotels, meals, and other expenses at spills, and special travel request forms that enable him to climb on an airplane no matter where he is when he is needed at a spill. The man can fly into the nearest airport, rent a car, and drive to the scene without bothering with the piles of paperwork that men in other commands might have to fill out to go the same distance.

The men are restricted, however, by the beepers they wear on their belts and keep at their bedsides whenever they are not at headquarters. The beepers can go off at any time to summon them to the office for another job. One morning Pete Brunk was getting on an airplane at Norfolk to attend an oil spill meeting in Milwaukee when his pager went off. He told the airline people to hold the plane and ran to call his office. A ship had run aground near Baltimore, on the Chesapeake Bay, he was told. He ran back to the airplane, got his luggage, and ran onto another airplane just about to leave for Baltimore. A Coast Guard helicopter had just landed at the Baltimore airport when Brunk arrived. It picked him up and plopped him down on the grounded ship forty minutes after the beeper sounded. He was there even before the local Coast Guard officials who had requested the Strike Team's help.

The beepers are not always that effective, but they do help insure that the men meet their readiness goal of getting at least four men and some Adapts gear airborne within two hours of an oil spill alert, and then augmenting those men within twelve hours with whatever team strength is required at the scene.

For his work in building the Strike Team, Chambers was awarded a Meritorious Service Medal early in 1976. It was an honor usually reserved for officers at the captain level. Chambers, a lieutenant commander, was two levels below captain. In the summer of 1976, he was named commanding officer of the Atlantic Strike Team.

In the three and a half years between Chambers' arrival in Elizabeth City in 1973 and the day the *Argo Merchant* went aground, the Atlantic Strike Team had reported to more than sixty major spills. The worst one for the team itself was in 1974, on the St. Lawrence River, where the tanker *Imperial Sarnia* had driven herself aground on a rock ledge, ripping open her belly and spilling 100,000 gallons of oil. During the salvage effort, Chambers and Dennis Perry, another diver on the team, were in the water alongside a support barge that was moored to the tanker itself. They were preparing to go under the ship and inspect the damage to the hull. Perry had been in the Coast Guard for six months, having joined specifically to get on the Strike Team. He had previously been in the Navy, where he was an expert deep sea diver. Both men were struggling with separate lines they were going to stretch under the hull to help mark the progress of their inspection when Chambers saw his companion drop below the surface. He figured Perry was trying to get out of choppy water. When he didn't bob back up within ten or fifteen seconds, Chambers looked up at the barge where the stand-by diver, with tanks on, was looking over the side.

"Where did Perry go?" Chambers asked.

"He went under the barge."

"Well, get in the water. Let's go get him."

The diver jumped in and the two of them went looking for Perry. At a depth of sixty feet, the pressure on the stand-by diver's ears forced him to return to the surface. Chambers went on alone, going down as far as one hundred feet, until he was so low on air he had to give up as well. Perry was gone. No one ever saw him again. The team searched for his body for three weeks in the waters which dropped from forty feet on the ledge on one side of the tanker to two hundred feet on the other side. There had been no sign he was in trouble, no indication what had happened. Perry is the only man the Strike Team has lost. A monument in his honor stands at the team's headquarters in Elizabeth City.

The most difficult spill was in San Juan over Christmas and New Year's Day, 1975–1976, almost exactly one year before the *Argo Merchant*. A barge with 500,000 gallons of heavy fuel broke loose from its tug in heavy surf and slammed into a reef about four thousand feet offshore. Waves pounded the barge constantly, rushing in from deep water, then building up to fifteen feet on the reef before crashing into the barge. In the midst of their efforts to prevent oil from spilling, Chambers flipped over in a rubber Zodiac raft and tore a cartilege in his right knee. For the remaining weeks of the thirty-two days they were on that job, Chambers' men took him out to the barge in the rubber boat and lifted him aboard, where he sat with his feet propped up and supervised the operation, giving directions with his cane. The men worked eighteen hours a day, struggling against the surf and the reef. They finally dragged the barge into shore with power winches. They lost about 300,000 gallons and pumped out the rest.

Not all of the spills were so difficult. In fact, the day before the team was called to Cape Cod, Chambers and ten of his men were in Quantico, Virginia, where the barge *Elk River* had gone aground in the Potomac River. Just a tiny sheen of oil trailed from the barge as the owners pumped its cargo into a second barge. The Strike Team surveyed the shoreline and stood by with Adapts gear in case the barge's pumps, which continued to work after the grounding, broke down. The barge floated free about three in the morning on December 14. Chambers and Jim Klinefelter, dressed in wetsuits and diving tanks, broke ice as they jumped into the river to inspect the hull. Within ninety minutes, they found the source of the leak, a hole so small that Chambers plugged it up by jamming a pencil into it. Then he and Klinefelter got out of the water. The team folded up their shop and went home.

In those three and one half years, while working on spills ranging from the *Metula* to the *Elk River*, the Strike Team had pumped or cleaned up more than 160 million gallons of oil, Chambers estimated. That was enough to fill the *Argo Merchant* more than twenty times. And of the more than sixty incidents they had responded to, not one had been a failure.

As Chambers scrubbed the oil from his face and arms and hands early Friday morning and headed for his first full night of sleep in four days, he thought about the eerie whistles and the flooded engine room out on the tanker he had been forced to leave a few hours earlier. The *Argo Merchant* held another seven and a half million gallons for him and his team to pump. But she couldn't hold onto it much longer, he thought, not in the condition she was in as he left. The Strike Team could be foiled this time, without even another chance to get aboard her. Chambers thought the *Argo Merchant* would be gone by morning.

11

The Plan

At daybreak, Ian Cruickshank took the *Vigilant* in for a close look at what the storm that had passed through during the night had done to the *Argo Merchant*. He didn't have to look long. The ship was still intact, but a stream of oil 150 feet wide and three inches thick in places trailed from the vessel like blood flowing from a wound. Though the winds and seas were out of the northwest, the oil rode southwesterly currents away from the ship and ranged as far as Cruickshank could see. Thursday, it had been only an oily sheen, mostly washed out with the Adapts discharge, that rotary currents scouring Fishing Rip had spun out in a two- to four-mile radius around the tanker. Now it was an oil slick.

As Cruickshank moved in as close as he dared through the rain and heaving eight- to ten-foot seas, he saw what he thought was the source of most of that oil. Thick black geysers shot out of ullage openings like Roman fountains, showering the blackened deck with oil, which was then washed into the sea by breakers that buried the vessel's midsection.

The heading of the tanker had begun to shift counterclockwise during the night, from a northwest direction to almost due west, in the face of twenty-five to thirty-knot winds and heavy swells. Despite its hard grounding amidships, the vessel was pivoting on its sandy perch, its bow too buoyant and too lightly aground to stand firm against the relentless onslaught.

Standing nearby were the tug *Sheila Moran* and the tug *Marjorie B. McCallister* with the barge. Both had arrived just hours too late to be of any use in the effort to save the ship the night before. Now, as they tossed about in the swells, the empty barge yanking hard on its hawser against the tug, they waited. For the first time since the ordeal had begun forty-eight hours before, there was a lull in the action.

The abandonment the previous night had caused the Coast Guard to change its strategy in its fight against the tanker. It had also changed the tempo. Since all the passengers finally had been evacuated, it was no longer a search and rescue mission; it was, as Captain Hein and Captain Folger had assumed from the beginning that it would be, a pollution incident. In fact, the oil was now the only major consideration. The Coast Guard declared that morning that the ship itself had deteriorated too much for it to be saved. The goal now was just to pump the oil out of her.

But success in removing the oil from the *Argo Merchant*'s tanks was unavoidably linked to the life and condition of the tanker itself. The Strike Team's only hope of preventing a massive oil spill was to keep the ship together. In theory, they were out to save the oil. In practice, they had to salvage the ship as well.

Just how stable the tanker was that morning was difficult to assess. That it had managed to survive the night was a relief to some Coast Guard officials on Cape Cod and in Boston. It was a surprise to Chambers. And no one was willing to predict how long she would last.

"The ship could sit out there for fifty years," Chambers told news reporters at the air station that morning. "Or a storm could come up and break it up tonight."

With the stern much deeper in the water, however, and with many of the thirty tanks open to the sea, the *Argo Merchant* was beyond Wednesday's seat-of-the-pants appraisal that if the vessel were to be saved, the engine room would have to be pumped out. It was now a major salvage job, which demanded careful planning of strategy and a fast marshaling of marine experts and heavy equipment needed to pump and contain the oil. There would be no more jumping out of helicopters in the middle of the night over a black ship, no more fretting over midnight deliveries, no more arguing with the master over tying a line to the vessel. The extraneous matters were cropped from the picture and the task reduced to its essentials: the Coast Guard would do battle against the sea and the weather to determine who would get the oil out of the *Argo Merchant* first.

The lull in the action over and near the tanker passed quickly. By 9 A.M., helicopters had airlifted four remaining *Argo Merchant* crewmen from the *Sherman* to Nantucket and transferred two scientists and a member of the Strike Team from the *Sherman* to the *Vigilant*. With the search and rescue mission over, the large Coast Guard cutter was no longer needed. At 9:05 A.M. Captain Veillette was released to return to Boston. He and his crew would be home for Christmas. Actual control of the incident passed officially to Captain Hein, while Cruickshank became the commanding officer at the scene of the grounding, to work for Hein.

The *Vigilant*, which had already provided a damage control team, drawn a detailed bottom profile of the shoal within two miles of the tanker, served as a refueling depot for the small helicopters, and served lunch to the Strike Team, was now becoming a scientific outpost as well. Oil spill scientists armed with current meters and drift cards were now among the variety of people who had gathered at the air station and were taking helicopter shuttles out to the scene. Their mission was to monitor any oil that spilled from the tanker and predict whether it would head for shore. They were also trying to gather some real-life information to plug into computer formulas that had been designed to predict the flow of oil on water on the basis of hypothetical wind and current directions.

In making his tours around the *Argo Merchant*, Cruickshank had picked up some raw current information himself; at one point, the *Vigilant's* propeller was turning for a speed of eight knots and the vessel was sitting still in the water, just holding her own against the flow. He had also learned that the tidal current around the shoals did not run back and forth like the ebb and flow on a beach. Instead, it rotated clockwise around the compass, coming from virtually every direction during one tidal cycle, and leaving almost no slack tide. The current was constant, and flowing past the tanker at speeds of three to five knots.

While a handful of scientists made their observations from the deck of the *Vigilant,* Joe Deaver, an aerial oceanographer, prepared to take a look at both the slick and the ocean currents from the air. Deaver, a civilian Coast Guard employee, spends most of his time charting the movement of ocean currents and ocean fronts much the way a weatherman keeps track of the movements of air. Using instruments that measure the ocean's infrared radiation, Deaver can fly over the water and accurately determine temperature differences between water masses within six-tenths of a degree Celsius. Deaver could fly along a beach in the summer and detect precisely where effluent from a sewage outfall hit the water and then determine how far it extended from the pipe before it was absorbed by the sea water. He also had flown over nuclear power plants to determine the area affected by their discharge of cooling water into the sea.

Deaver had been flying a routine current mapping route Thursday when he landed by chance at the Cape Cod Air Station. He had planned to catch a commercial flight back to Washington for Christmas. But as his flight made its final approach to the air station, he received a radio message to call his Washington office as soon as he was on the ground. His services had already been offered to Captain Folger and Captain Hein.

In the open ocean, the border between two water masses of different temperature becomes a current, just as the border between two air masses becomes wind. Scientists believe that currents are critical factors in the movement of oil on the water; thus knowing where they were off Nantucket would be indispensable to predicting whether oil escaping from the tanker would head toward the coast or out to sea. Deaver and his infrared gadgetry could track oil and locate the currents at the same time. The advance warning his information could provide might give Hein and pollution control workers on shore extra time to protect sensitive environmental areas against the *Argo Merchant*'s cargo.

At 11:45 A.M., Friday, Deaver took off in a Coast Guard airplane on the first tracking flight over the tanker's slick. It was snowing over the wreck, and the pilot had to maintain an altitude of five hundred feet to stay below the cloud cover so Deaver's instruments would work. It was as much an orientation flight for Deaver as it was a tracking mission. He needed to get a good picture of what he was dealing with, some bench mark information to which he could compare future flight data. The snow forced an early end to the mission, but he managed to locate the slick. It was a five-mile-long finger of oil, with a slight crook about two and one half miles out, and it pointed northwest—right at Nantucket.

Nantucket's Main Street, a quaint cobblestone plaza lined with wood-shingled shops, was lit up with Christmas trees, closed to traffic and alive with shoppers in the days just before the *Argo Merchant* captured the island's attention. A light dusting of snow had given the island landscape a look of Christmas card scenery, the kind of setting that "off-islanders," as mainland residents are called, imagine always exists in such places.

When news of the grounding first reached the island, many of the residents, hardened by the wreck-filled history of the shoals, reacted almost out of habit with a cynical shrug. But with the failure of the Coast Guard to refloat the ship in the first two days, they began to recognize that while the ship itself might be just another wreck, her cargo of oil would not go down with her. If the vicious currents and waves around the shoals brought the oil toward shore, it would threaten the island's livelihood. Nantucket has two industries: fishing, worth about one million dollars annually, and tourism, a fifteen-million-dollar business on which the island depends for survival. Seven and one half million gallons of oil could drown them both. With the news Friday that oil was leaking from the tanker and drifting toward Nantucket, islanders spent the day with one ear tuned to the radio.

Some residents had recently been preoccupied with oil anyway. The U. S. Department of Interior had held hearings in Boston and Providence, Rhode Island, the week before on its proposed sale of offshore drilling rights in the Georges Bank area to U. S. oil companies. Bill Klein, the twenty-nine-year-old director of the Nantucket Planning & Economic Development Commission, had attended the hearing just eight days before the *Argo Merchant* grounding to ask the federal government to prohibit offshore drilling within fifty miles of Nantucket. Now he heard his own words echoing in his ears. "It is likely that there will be a significantly higher reliance upon barges and tankers for transportation [of Georges Bank oil] ashore, rather than pipelines," he had said at the hearing. "Tanker and barge spill statistics clearly indicate that this form of transportation ashore is the most dangerous mode. Tough weather and Nantucket Shoals' notorious historical reputation for marine accidents will contribute to the situation by making areas within fifty miles off shore an especially high risk area as far as major spills are concerned."

The *Argo Merchant* had documented those words with more evidence than Klein had ever hoped to find. Now he and other islanders wondered whether something else he told the hearing panel would hold up as well. "Georges Bank and Nantucket Shoals are noted areas for rough seas and bad weather," he had said, "making spill containment nearly impossible a good portion of the time."

The stockpiling of oil spill equipment at the Cape Cod Air Station had actually begun with the Strike Team's arrival Wednesday afternoon. It had continued in earnest since then, as C-130s and a larger Air Force C-141 shuttled back and forth, day and night, between Cape Cod and Elizabeth City delivering gear. By dawn Friday, two Yokohama fenders, eight smaller fenders, a high seas oil skimmer, two more Adapts units, more than a half mile of high seas oil barrier, and an assortment of other Strike Team gear, including boats, rafts, tools, and diving equipment, were assembled at the air station. Most of it was piled up just off the runway.

Captain Hein had ordered the *Spar*, another Coast Guard buoytender, out of Portland, Maine, to the area to help the *Bittersweet* move equipment over the water. The Army had driven in trucks from Fort Devens, Massachusetts, to help move it over the ground. And the two huge Army Skycrane helicopters were due from Fort Eustis, Virginia, at noon, when they would be ready, if necessary, to move gear to the scene through the air.

Meanwhile, private pollution contractors, lured by news reports about the incident, were hauling oil spill boom of all shapes and sizes into the air station by the truckload, speculating that if oil came ashore, they would be in business. Pete Brunk, the Strike Team officer who worked nearly forty-eight hours unloading, loading, and reloading equipment from planes to trucks to boats and back to trucks again, had been working at spills for more than three years and he had never seen so much boom.

"Hell," he remarked, "we've got enough gear here to build another Panama Canal."

Early Friday morning, the *Bittersweet* had returned to Woods Hole from its Adapts delivery assignment, to pick up the Yokohama fenders, turn around, and head back to the tanker. However, anticipating that the storm ravaging the crippled vessel might open her up, Hein decided the buoytender should pick up two boxes of high seas containment boom instead.

While the Adapts pump is the Strike Team's key oil spill prevention tool, the high seas boom and a high seas oil skimmer are the nucleus of the spill containment gear the team maintains at its headquarters in Elizabeth City. The result of nearly a decade of research and development spurred by the *Torrey Canyon* accident, the boom and skimmer are evidence of the progress in pollution cleanup technology since wood chips, straw, and chemical dispersants proved the only effective weapons against the *Torrey Canyon*'s cargo. The skimmer, which sits high on pontoons and looks designed for lunar exploration, is equipped with a special belt or disc that repels water and attracts oil as it rotates, licking up the oil contained by the boom and depositing it in a bin from where it can be sucked into vacuum trucks or pumped into barges. The boom itself comes in six-hundred-foot sections packed in boxes weighing fifteen-thousand pounds. Each length is a fence stretching between dozens of inflatable pontoons. When the release is tripped, the boom falls out of the box, then jumps up firm in the water as carbon dioxide cylinders fill the pontoons with gas. In twenty minutes, the entire six hundred feet sits up three to four feet out of the water, ready to be maneuvered into position. Several lengths can be strung together without fear of tearing or breaking it apart under the pressure of the sea or the oil behind it. Once the boom is no longer needed, it must be deflated and cleaned and carefully folded as it is put away. The carbon dioxide cannisters must be reactivated. Thus it loses some of the efficiency available in its twenty-minute deployment: it takes four men three days to put it back in the box.

But as the *Bittersweet* sailed back toward the *Argo Merchant* with 1,200 feet of the boom aboard, it appeared unlikely that it would be used that day if at all. Chambers says it is the best boom in the world. Yet it will not hold oil in a current of more than a knot and a half—no boom will—and it is ineffective in seas over five feet and winds over twenty knots. Weather and current conditions at the scene exceeded each of those limits that day. It would have to calm down substantially before the boom could be used and skimmers could be brought to the scene. The swift current seemed likely to preclude use of the boom alongside the tanker where the oil was leaking into the sea. And once the oil drifted away from the vessel, it was spreading out so that even several thousand feet of boom would not capture much of it. Prospects for getting the oil out of the water were not good. Hein sent the boom out as a contingency. The only reasonable chance he had of containing the spill, however, was to pump oil off the tanker before much more of it poured into the sea.

Out on the *Vigilant,* McKnight and the three other Strike Team members were eager to get back aboard the tanker to check her condition first hand and to see how their equipment had survived the night. During the morning tour around the ship, the *Vigilant* couldn't get close enough for a good assessment, and McKnight knew no one was going to do any work out there until someone got aboard to take tank soundings to see how much oil she had lost and to see how she was holding up to the sea. Chambers didn't plan to return to the scene that day, but McKnight and the others could make the necessary evaluation.

Early in the afternoon, McKnight asked Cruickshank if the air station could send out a helicopter to take them aboard. By 3 P.M., he and two others were hovering over the tanker. But this time they were lowered to the bow. It was the only safe landing spot on the ship. What wasn't covered with water on the main deck after the bridge was caked with oil.

When they got aboard, they took soundings in the forward tanks, on the bow. But the men couldn't get down on the main deck. They could barely traverse the catwalk between the two deck houses. McKnight watched as oil continued to spurt from ullage openings, coating everything with a two-inch layer of gunk. And he checked the Strike Team's equipment. The pump was still sitting in the number five starboard tank that had caused trouble the day before. The pallets with the unused Adapts equipment were still intact, though they were coated with oil. But the 1,200-pound prime mover that had powered the pump in the engine room was upside down in a corner near the afterhouse, taking a beating from the waves. During the night, the breakers had ripped it loose from its strappings, picked it up, and flipped it over. It was ruined.

The men also looked into the engine room, which they had to enter on the first deck above the main deck. The water and oil were more than forty feet deep in the same hold where just twenty-four hours before, the Adapts had held the level steady at fifteen feet. The best they could say for the engine room now was that it was still attached to the ship.

After an hour-long inspection, the men had done all they could and were ready to get off. But they weren't ready to report the soundings to the beach, not by radio. The three of them aboard the tanker decided that the best food, the best shower, and the best beer were ashore. Cruickshank's hospitality aside, they wanted to join the rest of the team at the air station, and they figured their only ticket there was the information they had just gleaned from the tanker. McKnight thought the air station might send out a small helicopter to pick them off the *Argo Merchant* and then land on the *Vigilant* for the night. He got on the radio to Cruickshank.

"Commander, instead of having them send a small helo out, why don't we get a big one out here to take us right to the beach," he said. "I imagine they're going to want us for a meeting in there. Captain Hein is going to be there and he'll probably want to talk to us. Maybe they should take us back to the beach."

"All right," Cruickshank replied. "I'll give 'em a call."

By 4:30, McKnight and the other three Strike Team members, including one who had stayed on the *Vigilant* during the inspection of the tanker, were on their way to the air station.

When the Coast Guard formally took over the *Argo Merchant* under the terms of the Intervention Convention, it also hired its own salvage master. In a one-step bureaucratic procedure through the superintendent of salvage office of the Navy, Captain Hein hired the Murphy Pacific Salvage Company, just as Chambers had predicted. To handle the job, Murphy Pacific, anticipating a major role in the grounding when the shipowners hired them, had already made a telephone call to a farm in Middlebury, Vermont, and recalled from semiretirement their senior salvage official, Captain Alfred Kirchoff. A crusty, hard-bitten man with a strong trace of German in his speech, Kirchoff has more than fifty years at sea behind him and he has salvaged more than 240 ships, including small oil tankers torpedoed by German submarines off the Atlantic Coast during World War II and the old Queen Elizabeth, which he towed to refuge after it had broken down at sea. What he doesn't know about ship salvage hasn't been thought up yet, and his skill at snatching wrecked ships from the jaws of the sea is respected around the world.

When a ship runs aground and starts taking on water, the laws of physics that make her buoyant enough to float are upset

The Plan

and begin working with the sea to flood and weaken the vessel and sink it. The job of the salvage master is to devise a plan and mobilize the equipment needed to overcome the effects of the flooding and get the laws of physics working in the ship's favor again before it succumbs to the sea. At 2:30 P.M., Friday, Kirchoff, Hein, and Chambers met with some other salvage people to devise such a plan.

Nine years before, Kirchoff had been faced with a similar problem. On March 7, 1968, the *General Colocotronis*, a Greek-registered tanker carrying 92,000 barrels of Venezuelan crude oil—about half as much as the *Argo Merchant's* load—ran aground on a reef in the Bahamas. The only way to save the vessel and the oil, Kirchoff determined, was to pump out her cargo tanks until she was light enough to be towed off the reef. He hired a small, empty tanker to maneuver in close to the stranded vessel. The second tanker gave him both a receptacle for the cargo and a supply of heat to warm it so it would pump easily from one ship to the other. Kirchoff got 87,500 barrels out of her—ninety-five percent of the cargo—before he ceased pumping operations, added buoyancy to the ship by filling some tanks with air, and then towed her off the reef.

Kirchoff proposed an almost identical plan to salvage the *Argo Merchant*. Using a second tanker, however, was out of the question. The shoals were too shallow and too treacherous. Any skipper who might have dared try it before wouldn't think of negotiating the shoals after seeing how the *Argo Merchant* had failed. Besides, Captain Hein already had the receptacles—two barges, one which had just left the scene for the lee of Nantucket to seek shelter from the pounding sea, and a second barge which was underway from New York City and due at the scene by noon the the next day.

But the barges could not supply steam to reheat the tanker's cargo and that was one of the key elements of the plan. The tanker had been without power for more than two days. With the heating coils shut down, the oil in the tanks was surrounded by air and sea temperatures between thirty and forty-five degrees. When the vessel hit the shoal, the cargo temperature was between 100 and 110 degrees. The men were certain it was already too cold to pump easily. And by the time the tanker would be rigged for pumping, the oil's consistency would be much closer to asphalt than fuel oil. Without heat, the Strike Team could shovel it off the ship almost as fast as the Adapts pumps could pump it. Kirchoff needed a steam boiler and generator taken to the scene aboard a large, ocean-going vessel.

With two barges standing by, there would be little delay in the pumping operations. Once the first barge was full, it could transport the oil to shore for discharge there while the second barge was moved into position alongside the tanker.

But sidling a four-hundred-foot barge up to the *Argo Merchant* for several days in rough seas and currents would probably give the salvage workers two wrecks to contend with instead of just one. They would need a mooring system, a string of four heavily anchored buoys set up just off the port side of the tanker for the barge and the supply boat to tie up to.

The men also wanted more pumping capacity so they could move as much oil as possible as fast as possible. They planned to connect heating lines to one large center tank then pump cold oil into it from four to six other tanks. That way, they hoped, they could keep enough oil in the warm tank for one or two pumps to operate constantly and they wouldn't have to stop to rerig the heating cables each time they emptied a tank. Chambers already had five Adapts at the air station or on the tanker. Murphy Pacific had a couple of similar systems of its own. And through the Navy, Captain Hein decided to obtain from a Detroit company two larger commercial pumps with twice the capacity of the Adapts.

Finally, Murphy Pacific's own salvage tug, the *Curb*, was underway from Florida, having set sail on Thursday when the firm first became involved in the incident. It was equipped with pumps and spare mooring buoys and was due at the scene of the grounding sometime Monday.

Kirchoff's plan was to place the mooring buoys off the port side of the tanker with Coast Guard buoytenders, then tie the barge to the buoys parallel to the tanker and lash it down to both the wreck and the buoys with power winches. He wanted the barge held twenty-five to fifty feet away from the tanker. The Yokohama fenders would be draped over the port side of the tanker to serve as bumpers. The vessel with the steam boiler, generator, air compressor, and extra pumps would tie up to a buoy at one end of the barge. Hoses and power lines would run from the supply boat to the tanker and discharge hoses would go from the tanker to the barge.

Once the scheme was set up, the pumping would begin according to a carefully calculated plan designed by a naval architect to relieve the vessel of the intense stress caused by the flooding and the grounding. But they also would have to prevent pumped out sections from creating unwanted buoyancy. Kirchoff wanted the tanker stable until the Strike Team completed the pumping. That meant replacing the oil removed from the tanks with seawater to hold the vessel down.

Finally, once enough oil was removed to refloat the ship, the team would push the water out of the tanks by pumping in air in a process known as "putting a tanker on a bubble." In effect, they would make a gigantic air mattress out of the ship, filling the compartments with air just as Kirchoff had done with the *General Colocotronis*. That would make it buoyant enough, they hoped, to tow it off its perch to a safe area where its remaining contents could be removed.

But Kirchoff had had a distinct advantage with the *General Colocotronis* that he didn't have with the *Argo Merchant*. The Greek vessel had been firmly aground, bow to stern. He didn't like what he saw in the *Argo Merchant's* position, with the flooded engine room hanging over so much water and the bow bouncing up and down off the bottom.

About 5 P.M., McKnight and the other Strike Team members who had just returned from the tanker came into the meeting and confirmed earlier reports with their eyewitness accounts that the ship was pivoting on the shoal. Kirchoff frowned. He couldn't have the tanker spinning around in the wind and waves while the Strike Team tried to rig hoses to a moored barge. Chambers and his men would have to stabilize the bow by running the ship's two anchors out off each side and dropping them.

Kirchoff had had another advantage in the *General Colocotronis* stranding: time. Although it had run aground on March 7, he didn't even get aboard for ten days because of rough weather. And he didn't get his equipment set up for pumping until April 1. But he still got 87,500 barrels—about three and a half million gallons—out of the tanks by April 12, thirty-six days after the grounding.

The men figured it would take at least ten days and probably thirty to pump out the *Argo Merchant*. Unless the weather settled down, they doubted the ship would last that long. The forecast for the next day was not promising: gale warnings were posted for the shoals. Seas were expected to build to twelve feet.

But the plan was ready. The Strike Team was set to begin work in the morning setting out the anchors to steady the bow. They hoped the weather would give them time to act.

12
The Hawsepipe

For the first time since the grounding itself, the weather over the *Argo Merchant* Saturday morning was bright and sunny. As is often the case on Nantucket Shoals, however, the clear skies were accompanied by strong winds—thirty to thirty-five knots—that piled up eight- to twelve-foot seas, the heaviest since the day of the grounding. And now, after lying four days on the shoal, the *Argo Merchant* had worked herself deeper into the sand and her stern had dipped further into the water. She put up little resistance to the barrage of waves smashing over her, burying her starboard side and sweeping over her deck. She was leaning with a fifteen degree starboard list, and the water level on the low side back by the after deckhouse had reached the second deck—more than thirty feet above the vessel's water line. Occasionally, a wave would break over her smokestack.

During the morning, Cruickshank replotted the tanker's position with his loran set, at the request of officials ashore. His calculations showed that she had moved six hundred yards south from where she was when the *Vigilant* had arrived Wednesday. The power of the waves and currents had not only forced the tanker to pivot on the shoal; it also had dragged her more than a quarter of a mile along the ocean bottom.

Cruickshank also thought the *Argo Merchant* was leaking more slowly that morning than she had been the day before.

The ullage caps were spouting less oil less often, possibly because the cargo had cooled and wouldn't squirt through as easily as it had Thursday night and Friday, and possibly because so much oil had already leaked from the tanks that they no longer contained enough cargo to overflow as the waves passed under the ship. The slick was moving southeasterly away from the tanker, and it was composed primarily of heavy patches two inches thick and up to ten feet in diameter. Once caught in the huge waves, much of the oil seemed to be whipped into the foam and dispersed, Cruickshank thought. He had heard radio and television interviews with scientists concerned about "an ecological disaster," yet he couldn't see that the oil was doing much of anything. He figured a fisherman a few miles downwind wouldn't have seen any oil at all.

But Cruickshank's perspective on the path of the oil was limited to within two or three miles of the tanker. Joe Deaver and two scientists from the National Oceanic and Atmospheric Administration who had reported to Cape Cod, were in the air shortly after 8:30 that morning. They found the oil concentrated in a seven-and-a-half-mile-long horseshoe one to two miles wide that extended northwesterly toward Nantucket, then curled back toward the ship itself. It no longer pointed at the island and was actually farther away from land than the day before. In fact, the slick didn't really point anywhere. It seemed to be drifting toward the southeast and east, away from the coast and toward the Georges Bank fishing grounds. Deaver also sighted several large, scattered patches of oil twenty-seven miles east of the ship.

Without adequate tank soundings, determining just how much oil had leaked from the *Argo Merchant* was little more than a guessing game. The Coast Guard really had no idea how many gallons were contained in the horseshoe of oil. But they suspected that at least 100,000 gallons had escaped into the sea, and that morning they declared it a "major spill."

Under the national contingency plan, coastal spills from 10,000 to 100,000 gallons are considered "medium," while anything over 100,000 gallons is called "major." Once a major spill had been declared, the contingency plan is automatically triggered; the appropriate regional response team is alerted and the Strike Team is notified. But in this case, all that already had happened by now, and the official designation of the spill, like the decision two days earlier to take over the vessel, was little more than a formality. There were some extra precautions taken, however. Captain Hein ordered the Brant Point Coast Guard Station to make twice daily inspections of Nantucket's beaches to look for traces of oil. And the Commonwealth of Massachusetts' Department of Water Pollution Control hired its own pollution contractors to protect the island's harbors and beaches. The reason for such action was twofold: political pressure had begun to mount from some state officials who felt helpless as the vessel's plight became more precarious; and while the wind was blowing offshore and seemed likely to take the oil with it, underwater currents in the area move toward shore during the winter, and if any of the oil was sinking, it could head toward the coast.

There was little else anyone could do that day. Four members of the Strike Team had hoped to go aboard to take soundings, but it was too rough. Chambers and Kirchoff, however, did fly over the tanker that afternoon and look down as walls of deep blue and white water cascaded over her deck.

"You want to go aboard?" the helicopter pilot asked Kirchoff, at least half in jest.

The salvage master smiled. "No, I don't see any necessity in going aboard. I don't think it would make much difference at this point, do you, Barry?"

Chambers smiled back. "No sir."

"Besides," Kirchoff added. "I've seen them before. I don't have to go aboard."

They passed over the tanker and stared at it for awhile, then headed back to the air station to refine the salvage strategy based on their observations. Despite Cruickshank's report that the vessel might have been dragged along the shoal, the salvage master was confident that it was structurally stable and would remain aground, at least for a while. The weather was still the major problem, he thought. He had once worked on a salvage job off the Delaware coast in which a stranded ship was battered about for thirty-one days before it finally broke apart. Kirchoff never had a chance on that one. Of the thirty-one days, the weather had allowed him and his crew exactly ten hours to work aboard the ship. He didn't think the weather was going to give the *Argo Merchant* thirty-one days on that shoal. And he knew he would need more than ten hours aboard to save her.

On Sunday, the weather finally gave them the break they had been looking for. The morning was crisp and clear, with unlimited visibility, temperatures in the low thirties, winds about ten knots, and seas of four feet.

"Weather conditions are ideal for operations out here today," Cruickshank radioed to shore as he surveyed the tanker's condition.

At the same time, the slick had grown considerably from the *Vigilant*'s vantage point since the day before. Oil streamed from the starboard side of the vessel in a swath almost as wide as the tanker itself, as if a huge superhighway extended from her out over the water. Cruickshank could see it easily for three miles. About a mile from the ship, the slick was three hundred yards wide and it turned to the northeast.

Ashore, Sunday newspapers all over New England and around the country carried dramatic page one pictures of the stricken tanker caught in the heavy seas of the day before. Stories highlighting the details of the attempt to save the ship and the extent of the oil pollution were accompanied by predictions from scientists that the oil could wreak environmental havoc on the shoreline and over the Georges Bank fishing grounds if much more escaped. Television stations, some with their own film and some with vivid video tape footage taken by Coast Guard photographers, tantalized viewers with the elements of a sea disaster story—the waves crashing over the tanker, sending wind-whipped spray high into the air. It was a big story anyway, and since it was happening in a normally quiet news period just before Christmas, little else of significance was going on to compete with it. The eyes of much of the country were focused on Cape Cod.

Chambers had that in mind as he met with Captain Hein

and Captain Kirchoff that morning to plan the day's operations. The Strike Team had two major tasks to accomplish—setting the anchors out and placing the Yokohama fenders alongside the tanker—before one of the barges could return to the scene and pumping operations could be set up. The fenders themselves could have waited another day, but Chambers knew the crew of the Skycrane helicopters had been sitting around since Friday morning with nothing to do. They were itching to go to work, and with the whole world watching, Chambers didn't want anyone sitting around too long. He suggested that the Skycranes bring the fenders out that afternoon.

Meanwhile, Captain Kirchoff planned to go to Woods Hole to supervise the rigging of the mooring systems aboard the *Bittersweet* and the *Spar*. And shortly before noon, he hired the *Calico Jack,* an offshore supply boat moored at Quonset Point, Rhode Island, and ordered it equipped with a boiler and generator needed to reheat the *Argo Merchant*'s cargo. Two large high-capacity pumps, bigger but not as portable as the Adapts, were on the way from Detroit, also to be put aboard the *Calico Jack*. And Murphy Pacific's own salvage vessel, the *Curb,* complete with its own pumps and mooring buoys, was due at the scene of the grounding by midnight the next day.

With the additional pumps, the *Calico Jack* and the mooring systems, plus the 70,000-barrel and the 40,000-barrel barges already standing by, the marshaling of the salvage equipment was running smoothly. At the same time, four more high-seas oil skimmers, maintained by the Navy, were flown into the air station Sunday, to bolster the pollution-fighting gear already stockpiled there.

It was early afternoon by the time Chambers and six of his men were lowered to the deck of the *Argo Merchant.* The seas were nearly calm. No breakers were crashing over the deck. For the first time since they had abandoned the ship Thursday night, the men could get back on the main deck. Chambers had two of his men inspect the condition of the cargo tanks and take soundings from them.

Another Coast Guard helicopter was right behind the

Strike Team, escorting one of the Army Skycranes, which was not equipped with instruments needed to navigate over water. The Skycrane's rounded cockpit hangs down lower than the cabin and tail, so the helicopter looked like a huge bird flying over the water with its back hunched up, as if it were searching for food and poised for a dive.

The two five thousand-pound fenders are shaped like giant footballs, rounded instead of pointed at the ends and covered with dozens of truck tires that are strapped to them with heavy wire. On the ground, they look like a giant playground amusement on which a whole school full of youngsters could play "king of the mountain." They are called "Yokohama fenders" because they were developed in Japan for use as bumpers between fishing vessels, which go out to sea for months at a time, and supply boats, which come out periodically to deliver supplies to the fishermen and transport their catch to shore. Chambers had taken a lot of ribbing from other people in the Coast Guard for stockpiling the fenders, which are worth about $20,000 a piece. He didn't pay for them, however; he hadn't paid for much of his equipment. He had acquired it from other military units which no longer needed it. The Navy had used the fenders on some research project before Chambers had heard they were available. Now they were precisely what he needed to place between the tanker and the barge that would receive the oil.

Dangling beneath the huge helicopter, the two five thousand-pound mounds of rubber looked like eggs suspended on a string. And if they had been eggs, Chambers thought, the pilots and the crew of the Skycrane could not have handled them more delicately. Within a half hour, the fenders were floating in the water along the port side of the hull, tied to the tanker with six-inch double-braided nylon line the men had rigged to the fenders at the air station. The line had a breaking strength of more than 100,000 pounds.

Before the Skycrane left, it picked up one of the Strike Team's oil-covered pallets of Adapts gear that had never been unpacked since it had been hoisted aboard by the *Bittersweet*. It was the best chance they had, the men figured, of saving at least some of the gear that had been knocked around by the seas during the past two and a half days. Chambers wanted to check the gear out ashore and repair it if necessary so the team could bring it back out again to pump oil when the barge was alongside.

Veering the anchors, as the process is called, was a bit more complicated than setting out the fenders, and it required the assistance of the *Sheila Moran,* which had been waiting out the rough weather with the *Vigilant* since arriving Thursday night. Chambers planned to run a line from the anchor to the tug, which would then move away from the ship, pulling the anchor with it as the Strike Team paid out the anchor chain. He hoped to get each anchor out about one hundred yards at thirty degree angles off each side of the bow so they would serve as guy-lines to steady the ship and prevent the bow from pivoting and bouncing on the sand.

Chambers went down a ladder into a storage space under the fo'c'sle just forward of the first row of cargo tanks. He found a tangled pile of eight-inch manila line that had been used to moor the tanker in port. Nylon line would be stronger, he thought, and he returned to the deck.

"You seen any nylon line around?" he asked Darby.

"No sir, I haven't."

"Well, shit, we'll put this manila around the anchor then. I don't know if it's strong enough, but it's worth a try."

With McKnight and Joe Kuchin helping them, Darby and Chambers went back down into the fo'c'sle and dragged up two mooring lines, each about 150 feet long—enough, Chambers thought, for each anchor. The *Sheila Moran* could not move in close enough to the tanker to permit her crew to tie ropes to the anchors, so the Strike Team had to fasten lines to them from the deck, then heave the other end of the lines over to the tug.

Each anchor was tucked into a large notch in the hull just a few feet off the tip of the bow so that the arms and flukes that dig the anchor into the sand seemed to be mounted on the outside of the ship. The shank of the anchor extended up into a fifteen-foot hawsepipe, the channel through which the anchor chain passed when the anchor was dropped or raised. The shank was connected to the chain just below the top of the hawsepipe opening which poked through the deck about fifteen feet from the rail. The chain then ran out the remainder of the hawsepipe through a guide block to the anchor windlass, which turned to pay out the anchor chain or haul it in.

Chambers wanted to tie one end of the eight-inch line to the shackle where the anchor and chain were joined and run the other end down the hawsepipe to the anchor. Then, he thought he would reach over the rail with a boat hook, retrieve the line, pull the rest of it through the pipe, and toss it over to the tugboat. But when the men tried to snake the eight-inch line through the pipe, it got tangled up along the shank. And they found they didn't have a boat hook long enough to reach the anchor from the deck anyway. Thus the only way they were going to rig it was to drop the line over the rail and bring it up through the hawsepipe from the outside.

Chambers got a smaller, white line, tied one end to the eight-inch rope and the other to a pipe wrench. He dropped the wrench over the side and tried to land it on the anchor. But the pipe wrench kept bouncing off the flukes. Chambers devised a new plan. He made a beanbag by filling a pillow case he had found on board with some small metal shackles. With Kuchin holding onto him, he leaned over the port side rail and tried to swing the beanbag under the edge of the notch so it would plop down on the anchor. Even with the bean bag, the task was difficult. The wind was blowing about twenty knots now, and it whipped up under the bow, bellying out the white line and catching the beanbag in the gusts. The bag glanced off the anchor and clanked against the hull. Chambers pulled it up a few feet and tried again. It bounced off one of the flukes and then dangled free in the wind. Once more, he swung it back and forth like a pendulum, then tugged up on the line as the beanbag moved near the anchor. Once more, it knocked against the metal, then fell free.

"This is a pain in the ass," he remarked as he paused for a moment before trying again.

Finally, the beanbag landed with a dull jangle against the anchor and stayed there.

"Okay, we got the damn thing on there," he said. "Now let's pull it through."

One of the other men grabbed a boat hook and leaned as far as he could into the hawsepipe. He couldn't reach the beanbag.

"Well, screw it," Chambers said. He stepped back from the rail and looked down the pipe. Then he looked at Darby.

"Darb, how about you goin' down that pipe and getting' that goddamn line."

Darby examined the pipe, a hole about two feet in diameter already partially plugged with a link of chain and the anchor shank. It would be a tight fit.

"Yeah, I suppose I can get down in there," he said.

If any of them could, it was Darby. A thin, wiry five feet, six inches tall, he was the smallest man on the team. Chambers didn't even consider asking anyone else.

McKnight helped Darby tie a rope around his legs and held the end as Darby picked up the boat hook and crawled head first into the pipe, stretching his arms straight out in front of him. It was the only way he could fit. He boosted himself over the chain link, which was nearly half as big as Darby himself, then squirmed down the pipe, scraping his belly against the shank and pushing with his elbows off the sides of the pipe as McKnight gave him some slack.

"Okay, I'm doing fine," he said, his voice sounding like a muffled echo coming up the pipe. "Lower me down a little more. Okay. A little more."

As he worked his way down, he scratched bits of rust off the sides of the pipe. The wind whistled through, blowing the scrapings into his eyes. As he got close to the end, spray from a wave slapping against the hull blew into the pipe and splashed him in the face. He licked the salt from his lips.

"Okay, I'm at the end of the pipe."

He poked his head out and looked down on the heaving ocean a few feet below him. Then he looked over at the beanbag sitting on the anchor just out of his reach. He struggled to free his shoulder from the hole and stretched the boat hook toward the white line around the beanbag.

"I'm reaching it with the hook," he said, giving a constant progress report to the men at the other end of the pipe. "I've got it with the hook."

He slowly pulled the line toward him. "Okay, I've got the line. I've got a hold of it. He retreated into the pipe. "Okay, go ahead and pull me out."

He gripped the rope tightly in his fist and pushed off the bottom of the pipe to raise his stomach off the shank. As the men at the other end slowly dragged him out, Chambers fed him more slack over the rail. The beanbag dragged along over the shank.

Once Darby was out of the pipe, he helped McKnight pull the beanbag, the white line, and finally the eight-inch line out of it as well. When they had enough of the thick manila hawser to work with, Darby leaned back down the hole and tied the line to the shackle connecting the anchor and chain.

On command from Chambers, the skipper of the *Sheila Moran* backed in slowly under the tanker's bow, maneuvering carefully to avoid ramming it. Chambers had tied another white line, known as a messenger line, to the free end of the manila hawser and he tossed it over the rail to the stern of the tug below him. A crewman picked up the messenger, yanked it in quickly to pull the heavy line aboard before it fell to the water, then wrapped the eight-incher around a towing bit on the stern. The tug started to pull away.

Darby was up on the anchor windlass, waiting for instructions to start paying out the chain. He had released a stopper known as a cat's paw that clamped down over the chain and was keeping a brake on the windlass.

"Okay, Darb, let it run a little," Chambers said.

Darby eased off the brake. He and Chambers had to work to keep the line and chain to the tug taut so the anchor was suspended over the water. If it dropped, the strain of dragging it through the water would snap the line easily.

The chain clanked down the hawsepipe as the anchor slid from its perch in the notch under the bow.

"Slow it down, Darb."

Darby pressed down on the brake and the windlass slowed,

tightening the chain. The tug pulled the line taut. The anchor cleared free of the hull.

"Let it run easy."

He let up on the brake again. The chain ran off the windlass through the guide block and snaked down the hawsepipe.

"Slow it down again."

Darby leaned on the brake. The tug pulled the line taut. Out over the water, the rope strained against the chain and snapped. It whipped back against the tug. The anchor plunged into the sea with a deep, thunderous splash and went to the bottom.

Chambers grimaced. "Damn, that's not far enough," he said.

The anchor chain was out about one hundred feet, only one-third the distance Chambers wanted. And the anchor had taken so much chain to the bottom that it was probably no more than fifty feet away from the hull, barely far enough to provide any pressure against the bucking and pivoting. But it was in the water. They would have to do better on the other side.

The men lugged the other section of line up to the rail over the starboard anchor. Darby went directly down the hawsepipe as Chambers swung his bean bag up on the anchor. Darby grabbed the line with the boat hook and was hauled out of the pipe once again. He tied the eight-inch line off to the anchor shackle and went to his station at the windlass while Chambers tossed the other end of the hawser to the *Sheila Moran*.

The swift tidal current was running north-south through the rip, across the bow of the ship, which was on a westerly heading. On the port side, the tanker shielded the *Sheila Moran* from the force of the current and the tug had managed to point its bow into the northwesterly swells and pull the anchor out.

On the starboard side, however, the tug was sitting exposed to the full strength of the current. By the time the manila line was secured to the towing bit, the tug had slid past the anchor itself and was heading sideways into the seas. As the skipper tried to turn her out of the trough of the waves, the current pushed her back. The rope pulled taut across the edge of the hawsepipe.

"Be careful," Chambers told the skipper over the radio. "You're going to part that line on the—"

Crack! The line snapped with the sound of a cannon shot. It skimmed across the water and smashed against the side of the *Sheila Moran*. The Strike Team didn't even have a chance to release the anchor chain. The anchor hadn't budged.

Chambers looked over the side at the water swirling along the hull, then out at the broken line trailing into the water off the stern of the tug.

"Goddamn, that current is incredible," he said, shaking his head. "It just boils through here."

By now it was 4 P.M. The Strike Team wasn't going to work in darkness again, at least not until the pumps were set up and running.

"We'll give the starboard anchor another shot tomorrow," he told the *Sheila Moran's* skipper.

Then the Strike Team waited for a helicopter to take them back to the air station.

While Chambers and three of his men had worked on the anchors, three others secured hatches and valves, trying to make the tanker as watertight as possible and to prevent as much continued leakage of oil as they could. They also set up a radio on the bridge so they could speak directly with officials on the beach. They used it to report to them the results of the tank soundings.

The slick itself had more than doubled in size since the day before. During the day's tracking flight, Joe Deaver traced a ribbon-like plume of oil that extended westerly from the tanker on ocean currents for a few miles before curling back southeasterly and finally northeast in the shape of a sickle. It was sixteen miles long and two miles wide. The sickle was 90 to 95 percent covered with oil. But Deaver also had found lighter concentrations of the tanker's cargo within a pie-shaped segment of ocean that extended sixty-five miles east of the grounding. It was ten miles wide near the ship itself and thirty-five miles wide at its eastern edge. And it contained a lot more than the 100,000 gallons of oil Coast Guard officials had estimated had spilled by the previous day. When the Strike Team's soundings were compared to earlier measurements, officials recalculated the spill and determined that at least 20 percent of the *Argo Merchant's* cargo was gone. One and one-half million gallons of oil had already leaked into the sea.

During the first four days of the incident, the officials had expressed concern about the condition of the ship, but they stressed they were hopeful it would hold together long enough for them to pump out her cargo. On Sunday, however, Captain Hein and Rear Admiral Stewart, the first district commander, admitted that her condition was such that she could break up any time, particularly in heavy seas. They were preparing themselves for the worst.

Captain Hein told a news conference that afternoon that "if everything goes right," the mooring system and the barge would be set up in two or three days and they could begin pumping cargo within four to five days.

But there already were indications that everything wouldn't go just right. A storm was expected to hit the shoals in forty-eight hours, bringing winds of fifty to sixty miles an hour.

"We feel the vessel is fairly structurally sound right now," Stewart told reporters. "But frankly, it is my opinion that if that vessel is out there for an extended period of time and we have blows of fifty knots, there is a good possibility that she would break up."

13

The Current

The Strike Team returned to the *Argo Merchant* just before 8:30 Monday morning amid weather conditions even better than the previous day. The sun was shining brightly among scattered clouds and the wind was almost a pleasant breeze—blowing about ten knots from the southwest. The swells lapping against the ship were running two to three feet.

But the seas had been rough enough during the night to undo much of what the team had accomplished Sunday. As the men looked down on the ship from the helicopter, they could see that the Yokohama fenders were no longer floating in the water beside the hull. They were up on the deck, tangled in the ship's piping system. Overnight, the ocean had tossed the fenders around as if they were beach balls and had ripped the 100,000-pound strength line that had held them in place as if it were thread. The men would have to start over.

Once they were aboard, Chambers called to the shore to get the Skycrane back out there. While the team waited, a couple of them went below to find more line. Before the huge helicopter had brought the fenders out on Sunday, the team had rigged each one with two lines right at the air station where they could work on the bumpers while standing on the ground. Now they had to do it again, but on board this time, where just getting to the fenders was a struggle. The deck, the iron rails and catwalks, and the fenders themselves were covered with a two-inch layer of oil. The ship was so low in the water astern that even the light waves were washing over the cargo pipes and splashing the rubber hulks. The two men got the line and crawled and slithered along the catwalk, then fought for footing to brace themselves when they reached the fenders so they could cut and replace the tangled rope.

Meanwhile, Chambers, Darby, and two other team members went back to the bow for another try at the starboard anchor. This time, Chambers asked the skipper of the *Sheila Moran* if he wanted to use a wire cable instead of the eight-inch line. But the tugboat captain had already seen the line come whipping across the water at him and that was dangerous enough. If a piece of wire snapped, the damage to the tug could be devastating. And if anyone was standing in the way, the wire would slice him right in half. He wanted no part of wire. Chambers didn't blame him; he just wanted to mention the idea.

Chambers found some nylon line, stronger than the manila that had broken the previous day, and sent Darby down the hawsepipe again to attach it to the anchor. As Chambers prepared to toss the line to the tug, Darby assumed his position at the windlass. The crewman on the tug wrapped the line around the towing bit. The *Sheila Moran* groaned in the current as she slid past the anchor. Snap! The line pulled across the mouth of the hawsepipe and parted again.

"Son of a bitch," Chambers said, shaking his head half in frustration, half in awe at the power of the current. "Darb, we're gonna try it with two lines."

He got on the radio to Cruickshank. "Hey, Ian, how about sending your small boat over here and give us a hand rigging this anchor? The *Sheila Moran* just can't operate in this close."

Within thirty minutes, one of the *Vigilant*'s twenty-six-foot life boats, known as the *Vig-2*, was over the rail, in the water with a crew of men, and on the way to the *Argo Merchant*.

But by 11:30, while Chambers, Darby, and a couple other men tried to rig the anchor with two lines, the Skycrane arrived. The men who had scrambled about on the catwalk, their clothes now coated with oil, told Chambers the fenders were ready to go. The Skycrane lowered its hook to the deck. Two of the men slid around one of the fenders and attached the hook. Chambers came down off the bow to the front of the oily catwalk and watched as the fender rose off the deck.

This time, Chambers wanted the fenders further forward on the hull, where the ship was higher in the water. That way, he figured, they would rise up against the ship in the swells, then fall back as the waves passed without tearing the lines and getting thrown back on the deck. By 12:15, the two fenders were riding tail-to-nose in the water alongside the bridge house. Chambers was back on the bow.

Instead of using the *Vig-2* to rig the starboard anchor, Chambers decided to try to get the port anchor out a little farther. As the twenty-six-footer bobbed in the swells, Ensign Glenn Snyder, dressed in an oil-stained red wetsuit, struggled with the help of two other men to loop the eight-inch hawser around the chain. On the third try they did it, and passed the end to the *Sheila Moran*. The tug fought the waves and the currents and drew the line and the chain tight. The *Vig-2* couldn't clear out of the way fast enough, however, and got caught in the rope and chain. The strain was pulling the boat underwater. Snyder pulled out his knife and cut the hawser. The *Vig-2* lurched forward as if on a spring and smashed into the anchor chain. It was two hours before they succeeded in completing the job. The tug churned in the water and slowly moved off the port side as the Strike Team paid out the chain from the deck. They got the chain out another hundred feet before the line broke against the resistance of the anchor and the water.

Meanwhile, Chambers continued to work on the starboard side, trying to land the beanbag back on the anchor where Darby could reach it from the hawsepipe. With the starboard list to the tanker, the anchor was tilted over the water on a slight angle away from Chambers. Swinging the beanbag became a tiresome and frustrating job. Time and again, it glanced off the anchor or got caught in the wind. It was nearly 2 P.M. by the time Darby had gone down the hawsepipe twice to retrieve the beanbag and tie each line off to the anchor shackle.

The *Sheila Moran* backed in once again and picked up both lines as Chambers tossed them over the rail.

The crewmen moved quickly with lines as the tug's skipper fought the current along the starboard side. Darby was poised at the anchor windlass. But the tug got stuck in the trough again. She ran out all the line from the anchor as she slid aft on the current. She turned against the lines, bringing them tight across the hawsepipe, snapping them both. It was like stretching an elastic back as far as it would go. Both lines shot across the water and smashed into the side of the tug.

Chambers shook his head in amazement. "Incredible," he said quietly, a tone of resignation in his voice. He had worked on a lot of spills, but he had never seen a current like that.

"That's it," the tug's skipper said over the radio. "We'll never get it out of there unless we get a slack tide."

Chambers agreed. And there was no such thing on that rip. There was a period of slow water, perhaps, twice a day for about twenty minutes. But it certainly wasn't slow right now. They would have to try again the next day.

Captain Kirchoff had gone to Woods Hole Sunday afternoon to oversee the loading and rigging of the four mooring buoys he wanted placed in a line off the port side of the *Argo Merchant*. Each buoy system consisted of an 8,000-pound eel anchor, similar in shape to a ship's anchor, connected with ninety feet of two-and-one-half-inch chain to a 12,500-pound concrete sinker. The sinker was connected to the buoy itself with six hundred feet of one-and-one-quarter-inch wire. The sinker-anchor combination was specially suited to prevent the system from dragging. A 465-foot barge tossing about in the swells would put tremendous pressure on the buoys. With only an anchor holding the buoy down, the pressure on the chain would tend to pull the shank of the anchor up and eventually pull the anchor's flukes out of the sand, Kirchoff reasoned. With his system, however, the sinker would be likely to drag along the bottom, pulling the chain taut in the sand and pull the shank with it, flat against the bottom. The flukes would dig themselves deeper into the sand.

Two of the systems were hoisted aboard each of the buoytenders, the *Spar* and the *Bittersweet*, while they were in port at Woods Hole. The parts were carefully arranged in sequence on the forward section of each ship and then put together even before they left port. The process was hampered by the breakdown of the *Bittersweet*'s crane, which was essential for lifting the heavy anchors. Repairs took much of Sunday afternoon and evening. But by midnight, Captain Kirchoff sailed out of Woods Hole aboard the *Spar*. The *Bittersweet* left five hours later. Both vessels were on the scene shortly before noon.

During the morning, just before the *Bittersweet*'s arrival, Kirchoff scouted out the waters around the tanker and pinpointed the spots where he wanted the systems dropped—along a line about three hundred yards off the port side. The first would go at a forty-five degree angle off the bow, the next directly opposite the bow, followed by one just off the bridge and the fourth off the stern.

Aboard the *Bittersweet*, Lt. Cmdr. Jack Overath got his directions from Kirchoff about noon. Then, while the *Spar* set the first buoy, he ordered his men to prepare to drop the second. The crane operator lifted the sinker over the rail of the vessel at the buoy port, a wide gate off each side of the bow designed so buoys can easily be deployed or pulled aboard.

Overath maneuvered his vessel off the bow of the tanker, setting his distance by radar and taking a bearing to sail directly over the spot where Kirchoff wanted it.

"Let 'er go," he shouted.

One of his men cut the rope to the anchor. At the same time, the sinker was released, and tore cable from the coil as it dove for the bottom. The men watched the cable carefully as it paid off the deck, staying clear of what could be a lethal whip. Just before it was completely out, the crane operator lowered the buoy into the water and let it go. More than 22,000 pounds of gear had been stripped from the *Bittersweet* in less than twenty seconds.

The crew moved the parts for the second system into place, a deliberate process that took nearly an hour. Then Overath moved into position once more. The men cut a rope, pulled a stopper, and released a hook. The third buoy was in the water. The *Spar* placed the fourth, off the stern. By 2:15, the four buoys were in the water, ready for a barge.

Having given up on the starboard anchor for the day, the Strike Team still had a couple of chores to take care of before leaving the ship. The tightening of valves and hatches the day before had not succeeded in preventing oil from oozing out of the ullage caps, although it was leaking out more slowly now. But the team came up with another idea, hoping to shut off the leaks on the deck almost completely. Lt. John Clay, the team's third ranking officer, found some twelve-foot lengths of four-inch pipe below deck, just the size that would fit into the ullage holes. He and another member of the team went around the ship, jamming the pipes into the openings, lowering them about four feet into the tank, stuffing rags around the lip of the holes and lashing the pipes in as tightly as possible. Eight feet of pipe protruded from the holes. Instead of oozing out on deck when a wave smashed into the vessel, the oil now would be forced into the pipes, the men figured, where it would move up and down like liquid in a straw. It would take some rough weather, they thought, especially with the tanks now partially empty, to push the oil high enough to flow out of the pipes.

Meanwhile, the *Vig-2* had helped the team rerig and secure the Yokohama fenders. In the rush to reposition the two fenders with the Skycrane, the men had brought the lines from each one through the same mooring chock, the fitting through which rope passes from the deck to the outside of the ship, and wrapped them around the same bit. It had not been done in a very seaman-like manner, Chambers thought as he inspected the arrangement. And just as they had with the anchors, the men had underestimated the strength of the current. Over the past two and one-half hours, the strain of the fenders pulling in the current had drawn the wet lines up so tight that the men could not untie them.

To complete the rigging, Chambers wanted to move the front fender forward. If he cut its rear line, he figured, the fender would pivot forward in the current and his men could tie it off farther up along the hull. The Strike Team rigged a system to catch the line before it hit the water so they could pull it in. Then Chambers asked Clay to cut it. Clay stared for a moment at the tangle of lines wrapped around the mooring bit and tried to unravel them in his mind. Then he pointed to one of them and asked Jim ("Klink") Klinefelter to cut it. Klinefelter hesitated, then took his knife and sawed through the line. It snapped free. The front fender stayed right where it was. A moment later, the rear fender broke loose and drifted away from the tanker as Chambers and Clay looked on in disbelief. Clay had picked the wrong line.

Chambers got on the radio to the *Sheila Moran* and asked the skipper to retrieve the fender. Then he looked once more at the mound of rubber covered with truck tires floating out to sea and started to laugh. No job would be complete without one good blunder, he thought. Clay got the honors this time for cutting the wrong line. No harm was done. It was hardly a critical moment in the operation. And the *Sheila Moran* fetched it quickly. Instead of bringing it back to the tanker, however, the skipper towed it over to one of the mooring buoys and, with the help of the *Vig-2*, tied it off there. Chambers looked at it and figured it was better out at the buoy anyway. It wouldn't be thrown up on the tanker's deck again, he thought. And the fender would be just as effective draped over the barge as it would over the tanker.

By now it was 3:30. With less than an hour of daylight left, the Strike Team could get little more accomplished. Chambers summoned a helicopter from the beach. While they waited for the ride home, Klinefelter and Darby thought the ship needed a little sprucing up for Christmas. They most likely would be out there working Christmas Day, they thought, and if they couldn't enjoy the holiday at home, it would at least be fitting to have some of the trappings of the season on board. While scrounging around in holds for gear, the men had found two plastic Christmas trees, which Darby and Klinefelter propped up on either end of the flying bridge. And they also found the vessel's signal flags. Darby had already shinnied up the mast earlier in the day to rig a navigation light at the top to give other navigators some warning of where the darkened ship lay. Now he and Klinefelter went up again, to arrange a series of flags on the halyard stretched over the bridge. It was a greeting in semaphore: the flags read "Merry Xmas."

But Klinefelter wanted to make sure landlubbers and the people back home in Elizabeth City got the message as well, and Chambers consented to a little art work. Klinefelter found the paint locker below deck and broke it open. It was nearly full of water. He waded in anyway and sloshed around until he found a can of paint suitable for the holiday—deck red. He picked up a roller and a roller pad and headed back up a ladder to the bridge.

The paint had water in it as well, but Klinefelter didn't expect his work would have an artist's touch anyway. To accomplish his task of painting a message on the front of the bridge, he hung off the flying bridge, poked his head out portholes and walked down on a first deck above the tanks and reached up with the roller. He finished before the helicopter arrived, stepped back along the bow and surveyed the work. He nodded contentedly. It looked good, he thought. The message emblazoned in red: "Have a Merry. From Strike Team."

Klinefelter, a cocky, wiry blond-haired kid, knew his greeting might evoke frowns from the Coast Guard hierarchy who would not appreciate a light touch to such a serious event. But he didn't care. As he painted, he prepared a speech for anyone who might dare mention it to him. "Yeah, you'll have your fat ass here on the beach for Christmas," he was going to say. "We'll be out there working. Think about it. This is morale. You've got to have morale, you know."

Light southwest winds had blown most of the day, carrying the oil slick northeastward at an almost leisurely pace. Tracking flights found the heaviest concentration in a three-and-a-half-mile-wide banana-shaped section of ocean that extended sixteen miles to the northeast of the ship. A sheen of oil extended another seven miles to the southeastern edge of the slick for its full sixteen-mile length. Coast Guard officials made no revision in the previous day's estimate that one and a half million gallons had escaped from the tanker. But because of the relatively calm seas and because they had already found that the starboard tanks now had more water than oil in them, the officials guessed that the rate of leakage was slow, perhaps five hundred gallons an hour.

By the time Chambers and his men were back at the air station, the warm, sunny day had quickly turned into an overcast, dark late afternoon. The winds had picked up steadily during the day and the barometer had fallen. Predictions that a storm would hit the shoals in the morning seemed more accurate hour by hour.

The mood at the air station when the Strike Team returned was a mix of excitement and anxiety. Captain Hein was particularly excited. He had conceded to reporters at a late afternoon news conference that the weather forecast could hamper operations at the tanker the next day.

"It could be a very big problem, a constant problem," he had said. "But with two more good days, we would be in a position to pump oil."

And that was what he was excited about. Despite the frustrations the Strike Team had suffered with the anchors and the fenders, one anchor was out. The fenders were in place. Most important, the four mooring buoys were in the water. And the barges that would tie up to them were waiting in sheltered waters for the word to proceed to the scene. It had been a cumbersome machine that had worked to bring all this equipment together, Hein thought, and it had worked quickly and smoothly. All that he needed now was a supply of heat. In Providence, the *Calico Jack* was being rigged with the heating system. It would be at Woods Hole at daybreak to pick up more equipment and head for the tanker. He needed those two days. But the preparations were virtually complete. And even the dire forecast of fifty- to sixty-mile-an-hour winds for the shoals couldn't dampen Hein's spirits that night. The normally even-tempered and reserved captain was jubilant.

Chambers, however, was less optimistic. The small defeats that day at the tanker at the hands of the current had started to wear away his thick crust of cocky self-confidence, the feeling built up by an unbroken string of successes that he and his team and his gear could handle just about anything. The *Argo Merchant*, he thought, had gone astray in no ordinary type of water. It was a very special place that had refused thus far to give them any slack. He had already has his moment of jubilance that Thursday afternoon with his Adapts humming away and his other gear on the way and the feeling that he had the tanker by the balls. Then he had watched his moment disappear in the space of forty-five minutes.

Now, they had come this far. Kirchoff, Hein, and the rest of them along with his own men had assembled the gear and the manpower. They were ready to make their move. And he was ready to lead the charge. But he knew now that despite the work and skill that had brought them to this point, they had little more control over the outcome. It had come down to a matter of luck.

14

The Breakup

Pete Brunk, the strike Team's chief warrant officer, was at Woods Hole by 5:30 Tuesday morning to await the *Calico Jack*. Brunk was to help load more equipment onto the supply vessel, then climb aboard himself for the ride out to the scene. Shortly before 6 A.M., he called the Nantucket Lightship on the radio to get a weather report.

"Well, it's been blowing a little during the night," he was told. "But right now we've got about ten knots of wind and four to five foot seas."

"That ain't bad. That ain't bad at all," Brunk remarked. They could get their gear and the vessels out there and set up in those conditions, he thought. Maybe the weather would hold off for them.

"But the glass is low out here," the man on the lightship continued. "I think we're going to get a pretty good one out here."

Indeed, the glass, as the barometer is sometimes called, was low—about 29.10 inches and falling. It was a strong indication that if a storm hadn't hit there yet, one was on the way.

Brunk knew better than anyone else on the Strike Team, and probably better than any of the thousand other people who by now had gathered on Cape Cod and Nantucket to battle the spill, what a "pretty good one" was out on the shoals. He had

The Breakup

been the skipper of the Nantucket Lightship for eighteen months back in 1970 and 1971. A storm had hit the shoals on March 7, 1971. Brunk remembered the date because it was the day his father died. The Coast Guard had two 150-foot red-hulled ships with *Nantucket* written on the side in large white letters, and they alternated the sentry duty on the fringe of the shoals every few weeks. Brunk had arrived at the onset of the storm to relieve the other vessel. But for the next seven days it was so rough, that ship couldn't even pick up her anchor. Two lightships guarded the shoals that week, and maybe it was a good thing. The waves were so high that the ship's powerful beacons, perched on a mast fifty-five feet above the water, disappeared when the vessels were in the troughs of the swells. Though the picture of those waves stuck vividly in Brunk's mind, what he remembered most about that storm is that it was fourteen days before he could get off the lightship and fly home. He missed his father's funeral.

About 7 A.M., the *Calico Jack* steamed into Woods Hole. A huge forty-ton steel-gray boiler and two tractor trailer trucks full of equipment were welded to her deck. She was right on schedule.

But so was the storm. A winter gale out of the Ohio River Valley had moved easterly across southern New England and it hit Woods Hole shortly after seven. Winds gusted to fifty miles an hour.

At the air station, Chambers met with his team at 7:30 and picked the crew he wanted to take with him aboard the tanker. He briefed them on plans to try once more to get the starboard anchor out. The team also had to take soundings of the tanks so a final pump plan could be devised, and they had to prepare the ship for the arrival of the barge, the heating equipment, and the pumps.

The latest weather forecast for the area around Fishing Rip, issued two hours earlier, held little promise, however, that Chambers and his men would get aboard the *Argo Merchant* that morning. Thirty- to forty-five knot winds out of the west and northwest were expected to kick up five- to ten-foot seas by early afternoon and ten- to twenty-foot seas late in the day. Such a barrage would make work on the oil-slicked deck almost impossible.

Chambers was aware of the prediction, but he knew if he had worried much about previous forecasts, he would have done a lot of waiting for bad weather to come and his team would have accomplished little out at the tanker. In the five days he had been working on the ship, Chambers had learned not to depend on the weather out there, and not to depend on the forecasts for the weather out there. He met with Captain Hein and Captain Kirchoff under the assumption they would at least get something started that day. Chambers planned to take off from the air station by midmorning.

Meanwhile, on the *Vigilant,* Ian Cruickshank made his ritual early morning tour around the *Argo Merchant.* Southwesterly winds had blown twenty to thirty knots during much of the night, and though they had slacked off just before daybreak, they had picked up again. The stranded tanker was on a westerly heading and the swells out of the southwest, building up to more than ten feet, slammed against the bow on the port side. The whole front section of the ship was heaving up and down in the seas, and the deck from the bridge to the stern was buried in breakers. Spray ripped by the wind from the top of the waves showered the bridge and the smokestack. It was not a day for working on board, Cruickshank thought.

Once again, the sea had rejected much of the work the Strike Team had done the day before. The Yokohama fender that had been tied along the hull of the ship was back on the deck, having been tossed there during the night. The fender that had been tied off to one of the mooring buoys had also ripped loose. It was gone.

A thin layer of oil about one-quarter inch thick was drifting northeasterly off the stern, gradually breaking up into a sheen as it was whipped up in the pounding waves. While the slick was not as large near the ship as it had been a few days ago, Cruickshank thought oil was escaping at a faster rate than the day before.

"I'd suggest no operations on board today," he said, reporting his findings to the air station. "The only place you can think

The Breakup 203

of getting aboard is the bow, or maybe on top of the bridge. But the whole thing is awash, and it's tossing around pretty badly out here."

He ordered the *Vigilant* on what he calls a "chow course," pointing directly into the wind and moving ahead as slowly as the ship can go. The maneuver is designed to make a ride in rough weather as comfortable as possible during meals and to keep dishes full of food from tumbling off tables and splattering over the floor. Then he went below for some breakfast.

At 8:30, Cruickshank returned to the bridge. The *Vigilant*, still on a chow course, was about a mile away from the *Argo Merchant*. He walked out onto the bridge wing, just off the window-enclosed bridge itself. Standing at the rail in the brisk wind and freezing temperatures, he looked back for a moment at the embattled tanker that by now had sunk so low in the water astern, it seemed to be overwhelmed by the seas. The bow was pointed just slightly higher than it had been when they were in closer to the ship an hour ago, he thought. But it was a subtle change and Cruickshank did not stop to consider it. He ducked back out of the wind through the door to the bridge just as Ensign Glenn Snyder, who was on watch, squeezed past him to go outside for a look at the tanker himself.

By the time Snyder was at the rail, the vessel's bow pointed up sharply, at an angle of about twenty degrees. The midsection had sagged. He hurried back to the bridge, opened the door, and leaned in.

"Captain," he said. "I think she's breaking up."

Cruickshank peered through the rear window a moment, then walked quickly back out on the bridge wing.

"Oh yeah, it's going," he said, looking at the tanker. "It's definitely going. Let's get back there."

As the *Vigilant* turned in the swells, Cruickshank got on the radio to the air station.

"The *Argo Merchant* is breaking up," he said. "Tell Captain Hein the ship is breaking up."

Then he watched through the front windows of the bridge as the cutter bore down on the tanker. The bow was rising slowly, as if the ship were fighting the pain of a giant stepping on her back. She had opened up straight across her belly, just aft of the bridge.

After withstanding the relentless battering for six days while rocking and shifting on the sandy shoal, the *Argo Merchant* surrendered to the sea. The hull on the port side ripped apart under the stress. With the steel plating on the starboard side acting as a hinge, the breakers slowly pushed the forward section around, as if they were closing a jackknife. Oil gushed from the cargo tanks and stained the white water black as waves rushed across the broken hulk. It was ten minutes before the bow completed its swing and was lying side-by-side and grinding against the stern section, hanging from it like a huge broken gate.

For Cruickshank, it was like watching a slow-motion silent movie. He couldn't hear the scream of metal or the angry roar of the waves as they enveloped the ship. But the picture before him was vivid enough.

"Incredible," he murmured. "Just incredible."

Cruickshank's report swept quickly through the air station, leaving in its wake the weight of an emotional letdown. Until that moment, there had been a constant tension in the mood at the station, a sense of purpose that had built up during the past six days along with the sheer excitement of involvement in a major event. Now, within minutes, it was gone.

For Captain Hein, the report was a message of defeat. Everyone and everything had been ready. He had wanted just two more good days to set up and get started. Once in place, the equipment could withstand some bad weather, he thought. He didn't even get two hours. For a few silent moments, the effort now seemed so pointless.

Chambers sighed heavily and shook his head.

"Awww, shit," he said quietly. It was neither anger nor frustration. It was resignation. After yesterday's battle with the current, he had almost expected it would happen. It had been a week of small victories followed by small setbacks: a victory when the team landed on Cape Cod to find so many good decisions already made, and a setback when the first pilot was unable to deliver the Adapts; a victory when the pump was delivered, a setback when Chambers learned the other pumps were coming by boat. The latest victory had been the completion of preparations for pumping the oil. The logical setback was the cracking of the *Argo Merchant.*

But the more Chambers thought about the report that the ship had "broken in half," the more he wondered if it was really the final defeat. Since it had split aft of the bridge, he guessed that the break was around the number six and seven tanks, where the hull had started to bend under the weight of the water in the engine room the first night he was aboard. If that were the case, he reasoned, tanks one through five might still be intact.

"Maybe it's not a total loss," he suggested to Hein as they prepared to fly over the tanker to assess the damage. "Maybe it's still worth a try on the forward section."

By 10:30, two hours after the breakup, Chambers and Hein were hovering over the wrecked tanker. Forty-knot winds raised eight- to ten-foot seas that rolled toward the tanker, then piled up to fifteen feet as they hit the shoal, as if rearing back for a little extra strength before hurling themselves against the ship and driving the two pieces together.

Chambers had guessed right in estimating where the break had occurred. And while the tanks of the stern section were buried underwater, the first four rows on the bow could be reached, he thought. The nose of the ship pointed up in the air, bobbing in the water with the help of an air pocket in the fo'c'sle and an empty number one port tank. The bridge sat low in the water, putting the forward section on a steep incline.

"If we can get some of the buoyancy out of the bow and level her out a bit, we could pump those forward tanks out," Chambers suggested.

"Well, yeah, I guess it's the only chance we've got left," Hein replied.

When they returned to the air station and reviewed the ship's cargo plans, Chambers and Kirchoff figured they could level the forward section and maybe even refloat it by pumping the cargo from the number four center tank to the empty number one port tank. If that was successful, Hein figured they could seal the tanks, tow the hulk out to deep water well offshore, and sink it. If it didn't float free, they could bring one of the barges alongside, he thought, and empty as many tanks as necessary to refloat it and then sink the rest. Chambers briefed the team on the plan. He and Hein figured fifty to seventy-five percent of the cargo—5.7 million gallons—was lost. But that left nearly two million still within the ship.

"As long as we got five gallons of oil out there, we're going to try and get it," he said to the team. "I think there's still some oil in the forward tanks. A few of us will go out there first thing in the morning, rig up a pump, and try to move some cargo forward."

Some of the men, however, were not optimistic. The breakup had sucked their spirit out of them. It was a subdued meeting.

"How's our gear?" Darby asked. "Any chance of getting our gear off?"

Chambers shook his head. "I think we've lost everything we had out there. The whole after section except for the smokestack is underwater, and the water is just raging around there. The stern is a lost cause."

"What about getting some divers down there and see if we can recover some of that gear?"

"Well, I imagine we'll have to dive on her at some point to make sure there's no oil in the after tanks. Eight and nine and ten could have some oil in them, I suppose. Yeah, we can see if anything's left on the bottom. But nobody's going to be doing any diving with the water the way it is out there right now."

"You really think we can't get something out of the bow?" another member of the team asked.

"Man, I don't know. It's not in that good shape. But it's not a totally lost cause. It's the weather, you know, just the damn weather. We'll have to give it our best shot."

Darby, the eternal optimist, was ready to try. But he had already invested so much of himself in the effort that some of the spark within him died when the ship broke in half.

"I'll sure pitch in and give it a shot," he said. "But I don't know. We're just not getting a chance to work out there."

While Chambers worked much of the afternoon with Kirchoff, Hein, and the other salvage people, the rest of the Strike Team sat around in one of the buildings at the air station and watched a Coast Guard videotape of the *Argo Merchant* taken just after it had cracked apart. They watched in awe as the fifteen-foot breakers smashed against the broken hull, now helpless in the grip of the seas. They saw the raging white foam race over the deck and suddenly turn black. Over and over again, they played the tape. It left them speechless.

The barrage of waves that had split the tanker was also moving her spilled cargo. By midafternoon, the outer limits of the oil traced a rectangle to the east of the ship sixty miles long and twenty-five miles wide. It was not one massive slick, but an area of water blotched with hundreds of patches of oil known as "pancakes." Some were bigger than football fields. A thick patch of oil two to three miles wide extending six miles from the tanker was the heaviest concentration. The huge pancakes traveled thirty miles further to the east and fifteen miles north of the ship. Beyond that, the oil was much more widely scattered in the rectangle as it moved toward the southern tip of Georges Bank. It traveled slowly over the water, between one and one and one-half percent of the wind speed—about four-tenths of a mile an hour in a thirty-knot wind. Its progress and direction were influenced by tides and currents as well. So far, all the factors had worked to carry the spill away from the southern New England coast.

During the night, the wind increased in intensity. The men on the bridge of the *Vigilant* clocked it at forty-five to fifty miles an hour for ten consecutive hours, and they measured the waves in deeper water away from the tanker at twelve to fifteen feet. It was too dark to tell how high they were when they piled up on the shoal and battered the remains of the *Argo Merchant.*

At daybreak, Cruickshank moved in as closely as he could to assess the effects of the weather. The sun was shining and it glistened off the rails along the bow. The spray from the waves had frozen, coating the exposed parts of the ship with a sheet of ice. The bow itself was pitching violently in the seas and was raised much higher than it had been the day before. Cruickshank could see the beginning of the keel. There was one part, however, that he couldn't see. The bridge was completely underwater. It had disappeared.

With fifteen-foot seas running just off the shoals, Chambers knew he wasn't going to get aboard that day. Instead of taking a boarding team out there, he and Hein and Kirchoff and three other men took off in a helicopter about 9:15 to make another inspection of the ship.

Shortly before they left, Hein and Chambers were advised that they were to report to Boston that evening to testify before a Senate committee hearing called by Senator Edward M. Kennedy of Massachusetts to investigate the oil spill. The two men took a ribbing from some of the others on board, including the pilot and Pete Brunk.

"What are you going to tell them about how the oil will head straight for Hyannisport?" Brunk asked. "That ship could have broken right off the Kennedy compound." They laughed.

The ribbing and the laughter ceased abruptly, however, when they reached the ship just before 10 A.M. Kirchoff took a look at it as they hovered over it and shook his head.

"Forget it," he said. "She's gone."

The *Argo Merchant* had broken again. This time it was forward of the bridge, across the number three and number four tanks. The bridge itself had sunk. And the rest of the tanker's seven and a half million gallons of oil was pouring into the sea.

For the first time, Chambers had no comeback. There was no scheme left to try. No oil left to pump. Nothing left to call a ship. It was over. He looked down at the green-black water swirling around the pieces of the tanker and at the twenty-foot waves roaring over them.

The helicopter passed over the vessel a couple more times,

then turned away and headed back toward Cape Cod. The bow of the tanker receded into the distance until it was just a speck on the horizon. Then it faded out of view.

The men in the helicopter rode home in silence.

15

The Ship

As the oil gushed out of the broken tanker and spread over the ocean Tuesday and Wednesday, the anxiety and concern among those ashore turned to anger. Residents of Nantucket, who were especially vulnerable, were especially upset. Although they would feel the impact more than anyone else if the oil headed toward shore, they felt isolated from the decisions being made in Boston and on Cape Cod. No one on the island or in the Coast Guard had assumed a liaison role between the two, and the islanders learned mostly through news accounts what was happening on the mainland and at the scene twenty-seven miles off their shore. They turned their wrath on the Coast Guard.

"The Nantucket Board of Selectmen and County Commissioners would like to go on record as being opposed to the methods used in the unsuccessful attempts to remove the *Argo Merchant* and its cargo from Nantucket Shoals," the selectmen said in a telegram to Senator Kennedy. "We request a complete investigation of the U. S. Coast Guard's handling of this episode and further request an investigation as to why the vessel was so far off course."

The telegram spurred no such investigation, but as the islanders' anger subsided and was replaced by a feeling of helplessness against the whim of the wind and sea, there was a flurry of action by the federal and state governments and in courtrooms in Boston and New York.

Kennedy did, in fact, hold a hearing in Boston on the night of December 22, a few hours after Chambers, Hein, and Kirchoff flew over the remains of the tanker. Rather than investigate the Coast Guard's handling of the incident, however, Kennedy said, "We intend to understand, so far as humanly possible, why this tragedy occurred, a tragedy which is the nation's most devastating marine environmental disaster." He noted that the *Argo Merchant* represented only a fraction of the oil which is carried into Massachusetts ports by tanker. "Of the 6.3 million gallons of residual fuel oil that Massachusetts uses every day, nearly ninety percent is imported from foreign shores," he said. "The vast percentage of that oil is used during the winter months. We cannot afford this kind of Christmas nightmare ever again. We intend to be sure every precaution is taken which can reduce the chances of its recurrence."

Governor Michael Dukakis of Massachusetts and Governor Philip Noel of Rhode Island both sent telegrams to President Gerald Ford, urging him to declare New England eligible for federal disaster relief. And dozens of other political leaders called for tightening of standards for oil tankers and for stricter enforcement of standards that already exist. Congressman Gerry Studds, a Democrat who represents Cape Cod and Nantucket, renewed his call for oil spill liability legislation that would provide immediate compensation to anyone who suffered damages from oil spills.

The fishermen who earn their livelihood from Georges Bank were those most immediately affected by the oil spreading over the ocean. Fishermen on the wharves in Point Judith, Rhode Island, New Bedford, Massachusetts, and all over Cape Cod and Nantucket had followed the details of the salvage effort with concern throughout the week. They had just won a long battle in the Congress which would protect Georges Bank from overfishing by foreign fleets when it went into effect March 1, 1977. Now, just on the verge of a law that would help replenish the dwindling stocks of fish, a foreign oil tanker was threatening the fishing grounds with environmental destruction.

On the day before the *Argo Merchant* split apart, but after it had already leaked one and a half million gallons of oil into the sea, a fishing industry group headed by the Cape Cod Fishermen's Coalition filed a $60 million suit against the owners of the vessel and another $60 million suit against its master, Georgios Papadopoulos. A lawyer representing the Nantucket Conservation Foundation filed a suit on behalf of the Georges Bank itself, seeking money to restore any damages caused to the area by the oil and asking for higher construction standards for oil tankers entering New England waters.

After the vessel had broken and her cargo had bled into the sea, the Continental Insurance Company, which had insured the cargo, filed suit against the owner to collect $2,221,958.31— the value of the oil and the amount it paid out to the companies who may have owned it when the tanker ran aground. At the same time, Thebes Shipping, Inc., of Monrovia, Liberia, the owner of the *Argo Merchant*, went to court under an 1851 federal statute known as the Carriage of Goods By Sea Act, seeking to be exonerated from liability in the accident, or to have its liability limited to the value of the vessel and its cargo after the wreck. That value was zero. Shipowners seek limitation of liability under that law in virtually every marine accident in which damages are sought. Once filed, a limitation proceeding becomes the controlling legal action in a case and protects the vessel owners from any other lawsuits until the court determines whether the owners are indeed liable or not. To make that judgment, the court usually decides whether the accident occurred because of human error or because the vessel itself was unseaworthy on its final voyage. In a case of human error, the owners are freed of any financial responsibility for damages; in a case of an unseaworthy ship, the owners have to pay. Thus, the court's decision ultimately depends on its opinion of the condition of the ship.

As a Liberian tanker, the *Argo Merchant* flew a flag with one large white star in a field of blue and eleven red and white stripes. Its likeness to the American flag is particularly fitting in the shipping world, for it was a group of Americans, led by a former Secretary of State, who were responsible for establishing a tiny West African nation with no natural deepwater port as the home of the world's largest merchant shipping fleet and the world's largest tanker fleet. Known as a "flag of convenience" nation, Liberia has been a haven for thirty years for shipowners who want to reduce substantially their shipbuilding and labor costs, avert stringent safety standards and inspections, avoid American corporate taxes, and maintain anonymity.

It was in 1947 that Edward R. Stettinius, Jr., who was Secretary of State for about a year under Presidents Franklin D. Roosevelt and Harry S. Truman, called a meeting in New York of lawyers and politicians who eventually wrote Liberian maritime law, rewrote Liberian corporation laws, and formed the Liberia Company to exploit the nation's natural resources, particularly iron ore. Before he joined the State Department in 1944, Stettinius was president and chairman of United States Steel. The result of the efforts of these lawyers and politicians was a political and economic scheme that was especially favorable to producers of oil and steel and to the owners of vessels which shipped those products. It was also beneficial to Liberia. In 1949, Stavros Niarchos, a Greek ship owner, registered the first ship under the Liberian flag. Within ten years, 1,015 ships were Liberian, and by 1976 the size of the fleet had grown to more than 2,600. The government charges $1.20 a ton to register a ship in Liberia, and then ten cents a year per ton thereafter. The flow from these fees each year into the Liberian treasury represents about eight percent of the country's gross national product. Under the Liberian maritime and corporation laws, firms which operate out of the country are not required to pay Liberian taxes. Many of the ships are operated out of New York. If those vessels flew the American flag, their owners would pay 48 percent of their profits in taxes to the United States government.

The shipowners also circumvent U. S. laws which require that American flag vessels be constructed in the United States and manned by American crews. Even with government subsidies, the cost of building a tanker in an American shipyard is twice the cost of constructing the same ship in Japan, oil industry representatives say. The difference in salaries between American and foreign crews is even higher, and it is one of the most significant reasons why American ship owners flock to cheaper flags. The wages of a thirty-two-man American crew for an average 50,000-ton tanker in 1976 totaled $1.7 million. The same vessel flying the Liberian flag was manned for between $200,000 and $600,000, depending on the nationality of the crew. The lowest paid crews are usually a mix of nationalities, similar to the *Argo Merchant*'s crew, which was comprised of Greek officers and Pakistani, Filipino, Honduran, and Trinidadian crewmen, and which was paid about $275,000 a year.

Liberia is also an easy mark for sailors, no matter what their nationality or their qualifications. While, on paper, the standards for Liberian officers are similar to the rigid requirements for American officers, Liberia rarely applies them. Just as few real Liberian firms own ships registered there, few Liberians choose the sea for their livelihood. Thus most of the nation's licenses are merely handed out to foreign sailors who have obtained their ratings elsewhere and need an equivalent Liberian certificate to sail on a Liberian ship. Little, if any, effort is made to check the applicant's training or experience or even the authenticity of the foreign license. American seamen who disdain the arrangement say the easiest way to become an officer on a foreign flag ship is to steal an American license, photocopy it with the name blocked out, write in another name on the copy and photocopy it again. Then present the second copy to a Liberian official.

Registering a ship in Liberia is hardly more difficult, especially for American shipowners. The Liberian Corporate Services, located in New York, will help shipowners form Liberian corporations and register vessels within forty-eight hours with the Liberian Bureau of Maritime Affairs, located in the same office in New York. Financing can be worked out through the International Bank of Washington, which owns the International Trust of Liberia, another creation of the Stettinius group which runs the Liberian maritime operation. None of the key figures in the Liberian scheme is Liberian. They are American.

One critic of the arrangements has suggested that, as a nation, Liberia is a post office box.

Flags of convenience were not invented by Stettinius and his group. Since at least as far back as the fifteenth century, merchants from several nations have flown under foreign flags to avoid interference from countries at war with their homelands. More recently, some firms have sailed vessels under flags of nations where they have business interests so they can hire native crewmen. And during Prohibition, some American passenger liner operators found their ships were drier than their prospective passengers would like. To compete with foreign ocean liners, the Americans switched to foreign flags, so they could serve liquor on board.

Just before World War II, American merchants wary of the nation's uneasy neutrality toward the war in Europe sought the much more certain neutrality of the Panamanian flag to avoid troubles on the high seas. The switch continued during and especially after the war, when a flood of ships built in the midst of the war years hit the private market, sending their owners searching urgently for economically competitive corporate arrangements. Panama offered them and rose to preeminence among merchant shipping nations. In December, 1948, however, with the adoption of the proposals of the Stettinius group, Liberia embarked on a course that would end the Panamanian domination within a decade. Today, Liberia boasts by far the largest merchant marine fleet in the world.

The 1,029 oil tankers registered in Liberia in 1976 represented twenty-four percent of the world's tankers and thirty percent of the world's tanker tonnage. Among the companies that owned them were Exxon, Mobil Oil, Gulf Oil, Shell, the interests of Greek shipping magnates Aristotle Onassis and Stavros Niarchos, and Thebes Shipping, Inc., the owner of the *Argo Merchant.*

Most shipping observers agree that the fleet of ships managed by the major oil companies are among the most modern and best-manned vessels in the world. Exxon, for example, which sails about fifty of its 152 tankers under the Liberian flag and another twenty-four from Panama, hires and trains its own crews and keeps careful watch on their job performance, maintaining voyage-by-voyage records for each crewman.

Largely because of the modern fleets maintained by the oil companies, the average age of a Liberian tanker is six, compared to twelve for an American vessel. It is also the oil industry, not the Liberian government, that works hardest to insure that the modern fleet is properly equipped and manned by a trained crew. But most of the newer tankers are 100,000 deadweight tons and more, with drafts far too deep for the shallow ports of the United States. Yet about forty percent of the nation's oil imports in 1976 arrived by Liberian tanker. When the *Argo Merchant* ran aground in December, 1976, the daily imports into those ports equaled thirty-five *Argo Merchants* full of oil. The amount has increased substantially since. And the ships that bring it in are the smaller, older, and often marginally operated tankers, those which make up what is sometimes called the Second Liberian Fleet.

"You have two different types of fleets in that flag of convenience registry," observed Robert J. Blackwell, assistant secretary for marine affairs in the U. S. Maritime Administration in testimony before Congress. "Many are out to make the quickest buck, to put on the cheapest crew. They avoid maintaining their vessels. The fact is, you can't expect small countries which gather a great deal of their revenue from registering foreign ships or registering ships owned by other economic interests in their countries to exercise stringent control of these ships. They want that revenue. They want those fees. If Panama exercises strict control, the ships will run to Liberia. If the Liberians do it, they go to Malaysia. You just chase them around the world. They are looking for a place to rest where they can get the cheapest crews [and the] least stringent regulations."

But whether large or small, old or new, oil company-owned or independent, Liberian tankers, both in the year 1976 and over a thirteen-year period from 1964, had many more than their share of major accidents. In 1976, twenty tankers worldwide were total losses after maritime accidents. Ten of them—fifty percent—flew the Liberian flag, although only twenty-four percent of the world's tankers were registered in Liberia. Two of those casualties were supertankers of 224,000 deadweight tons and 275,000 deadweight tons, both less than four years old.

Another indicator of the Liberian performance is the percentage of tonnage lost over a thirteen-year period. According to statistics compiled by Arthur McKenzie, director of the Tanker Advisory Center in New York, Liberia ranks tenth among the fifteen largest maritime fleets in total tanker losses suffered from 1964 to 1976. Panama is eleventh. To determine the ranks, McKenzie divided the tonnage lost by the total tonnage in the fleet to come up with the loss percentage. The world tanker fleet lost .31 percent of its tonnage during the period. The Liberian fleet lost .50 percent. Panama lost .51 percent. Over the same time, the American fleet lost .15 percent, sixth in the world ranking behind Russia, West Germany, France, Japan, and Great Britain. Italy, with a loss of .64 percent, and Greece, with .76 percent, had the worst records of the fifteen nations in McKenzie's study.

In October, 1970, two Liberian tankers, the *Pacific Glory* and the *Allegro*, collided off the Isle of Wight in the English Channel. The *Pacific Glory* suffered an explosion that killed fourteen of her crew. Though both vessels were fully loaded, neither lost much oil. The *Allegro*'s third officer, who was on watch at the time of the accident, had no license, nor did two engineers on each of the ships. The outcry from the incident against flags of convenience prompted Liberia to set up a worldwide force of 150 inspectors to make annual boardings of each ship in the fleet, to check crew licenses, vessel documents, and navigational charts and equipment.

Liberian government officials have pointed with pride to the inspection program and held it up as an example of their effort to improve their fleet and shed the derogatory "flag of convenience" label. High officials of the United States Coast Gaurd have supported those statements. The Liberian officials also concede that about one hundred of their tankers are part of the "second fleet." In the five years the inspection program had been in existence through 1976, the *Argo Merchant* had suffered five significant casualties—including a grounding—that were reported in *Lloyd's List*, the daily newspaper of

Lloyd's of London. She had caused numerous minor oil spills as well. The Liberian government, however, had heard nothing about it. Its file on the *Argo Merchant* was clean.

But the tanker's clouded history went back far more than five years.

The *Argo Merchant* was built in Hamburg, Germany, in 1953 and first launched as the *Arcturus,* the name of one of the brightest stars in the sky. At 28,870 deadweight tons—her total capacity for cargo, supplies, and crew—she was a large tanker for her time, and her mission was to carry oil to Western Europe, which was switching its industrial plants from coal to oil. She was six hundred forty-one feet, four and three-quarter inches long and eighty-four feet, two and three-quarter inches across. She was forty-four feet, one inch deep from her main deck to her keel, and, when fully loaded, thirty-four feet, ten and one-quarter inches of her hull—about the height of a three-story building—were underwater. Her top speed was sixteen knots.

Little is known about her first eleven years at sea. If, by chance, she had had a record of flawless operation, it ended in July, 1964, when she damaged her propeller and cracked her propeller shaft, possibly by striking bottom. Exactly how or where the ship suffered the damage is not known, but it was the first of about twenty incidents that plagued the ship over the next twelve years before her grounding and death on Nantucket Shoals. During those years, she was sold like a used car four times to owners who either failed to care for her adequately or who supplied her with crews that neglected her.

When she damaged her propeller in July, 1964, the *Arcturus* was owned by the Fairplay Tanker Corporation of London and flew the Liberian flag. The propeller was replaced and the shaft was repaired in Palermo, Italy. Five months later, both steam turbine generators were damaged from excessive wear when the vessel arrived in Naples.

In August, 1965, the *Arcturus* was carrying a cargo of grain from New Orleans, Louisiana, to Yokohama, Japan, when she was forced to stop over four days at Cristobol, Panama, to repair the feed pump which brought water to the ship's boilers.

The tanker was sold to Bootes, S. A. in 1966 and managed by an agent known as Dammers & Van der Heide's Shipping and Trading Company, Limited. The vessel continued its registry in Liberia, but where the owner was located is not known. Whoever headed the company named it after the constellation Bootes of which Arcturus is the brightest star.

In April, 1967, the *Arcturus* collided with the *Tachikawa Maru* in Japan and went aground. Bow plating was torn, a gash was cut in the starboard side, and internal supports and holds in the bow were heavily damaged. No oil was spilled, and no damage was reported caused by the grounding. However, inspectors recommended that the hull, propeller, and rudder undergo a special examination at the vessel's next regular drydocking.

Reparis from the collision were completed in Singapore on May 14, 1967, and four days later the *Arcturus* left port, bound for the West Coast of the United States. Within two days out of port, three fires in the boilers damaged the ship's engines and the tanker was forced to drop anchor on May 20. Her propeller shaft was leaking as well, and sea water was coming into the bilges. A salvage tugboat from Singapore came alongside and had to provide steam, so the ship's crew could raise anchor, and pumps, so they could pump out the water in the bilges. The tug towed her back to Singapore. Repairs were completed by June 12, and the *Arcturus* left Singapore the following day, only to return again July 1 with machinery damage. She finally left Singapore July 10.

Two weeks later, off the coast of Japan, the *Arcturus* suffered another engine breakdown and had to be towed to Osaka for repair of one of the boilers, one of the feed pumps, and other engine room machinery. It took thirty-eight days. The tanker sailed again on September 4, bound for Los Angeles. Two weeks later, she lost power and was reported disabled six hundred miles from Midway Island. The Coast Guard provided fresh water and diesel fuel to the ship, which then limped toward Midway under her own power. By October 13, the tanker reached Honolulu, where surveyors determined the cause of the vessel's boiler troubles and subsequent power failures: improper boiler operations resulting from the negligence of the crew.

On November 17, she left Honolulu, only to have to return three days later because of minor electrical problems. It was December 2, nearly eight months from her initial departure from Singapore, before the *Arcturus* arrived in Los Angeles.

While entering a dock in Madras, India, in May, 1968, to discharge grain from the United States, the stern of the tanker slammed into a quay, causing extensive damage to the shell plating and some damage to the steering gear. The following month, also in Madras, the engine room flooded after both boilers and the emergency generator had broken down. The ship's crew refused to help remove the water. The flooding damaged electrical equipment in the engine room while the owner's agents hired salvage workers and looked for a new crew.

The *Arcturus* was eventually towed back to Singapore, where in mid-September surveyors found evidence of the damage caused by the mishaps of the previous two years, including the grounding off Japan in 1967 and the flooding damage in Madras. It was not until November 16, 1968 that repairs were completed. On December 9, the ship reported a starboard boiler drum had broken in four places. Bootes gave up on the *Arcturus* and sold her to the Republic of Indonesia, where she became part of the national tanker fleet and was renamed the *Permina Samudra III.*

On September 23, 1969, the *Permina Samudra III*, carrying 17,465 tons of crude oil, ran aground about three miles off Balikpapan, a seaport in central Indonesia. She was refloated thirty-three hours later with the help of tugboats. The master of the vessel said the bottom seemed muddy and there was no obvious damage sustained. He did say that he feared hidden damage in the hull and the engine room but he did not request an inspection. None was made and the tanker continued her voyage.

She came into Singapore on May 13, 1970 reporting machinery damage. In fact, the ship's entire power plant was in such disrepair that a surveyor found fourteen major deficiencies in a cursory inspection. Water and cargo pumps were broken, the steam system was dangerously contaminated with oil, the boilers were corroded or their tubes were broken, and the main turbine emergency gear, emergency alarms, and the engine room telegraph, which receives directions from the bridge, all did not work. The surveyor did not even inspect the cargo tank piping, which was reported in poor condition, because dangerous levels of gas remained in the tanks. He recommended a complete reconditioning of the whole power plant. By the beginning of June, 1970, the *Permina Samudra III*, unmanned, was under tow to a Japanese shipyard for repairs. But she would never sail again on her own power under that name. The vessel was sold to a Liberian company late in 1970, managed by the Trans-Ocean Steamship Agency in New York and renamed *Vari*.

On March 22, 1971, the tanker ran aground for the third time since 1967, this time off the coast of Calabria, Italy. She was in ballast and was refloated off the sandy bottom within sixty-two hours and then towed to port because her rudder was damaged in the stranding.

Sailing again in April, the *Vari* arrived in Marseilles, France, for repairs and surveyors found that her starboard turbine generator was heavily damaged because of boiler trouble. Other parts of the engine also needed overhaul. How many more voyages the tanker sailed as the *Vari* isn't known, but in 1973 she was sold once again, to Thebes Shipping, Inc., which named her the *Argo Merchant*.

Mark Madias, a stocky, bespectacled man with graying, receding hair and a long, full face, became chairman of Thebes Shipping, Inc., when it was formed in 1973 to buy the *Argo Merchant* for $2.4 million. He is also chairman of three other corporations, which own Thebes. These are the Argosea Shipping Company, the Olympian Shipping Company, and the Alcyon Shipping Company. Like Thebes, all are Liberian corporations and all serve as part of a complex corporate maze which hides the identities of the sixty to seventy stockholders in each of the three secondary corporations and which is intended to shield the stockholders from any liability in case of an accident.

Madias himself is a naturalized American of Greek descent. He owns between three and seven percent of the stock in the three owning corporations. Ninety to ninety-five percent of the other stockholders in each corporation are also Americans.

Madias is also chairman of Amership Agency, Inc., the New York-based general shipping agent for the *Argo Merchant* and for seven other vessels, all of which are owned by separate corporations and all of which have "Argo" in their names. These include the *Argo Castor,* the *Argo Leader,* the *Argo Master,* the *Argo Pollux,* the *Argo Trader,* the *Stolt Argo,* and the *Stolt Argobay.* (The *Argo Leader* has since been sold.) All are secondhand ships and all but one are at least fifteen years old. The other one was built in 1966.

Madias' brother Stelios is an owner of Triship Agency, in Piraeus, Greece, which hires the crews, especially officers, for all the vessels managed by Amership. Stelios also holds stock in the Olympian Shipping Company, one of the three second-line owners of the *Argo Merchant.* Mark Madias, however, does not have any financial interest in the Greek agency.

Nicholas T. K. ("Captain Nick") Skarvelis is president of Amership and owns a one-eighth interest in the company, as does his brother, James, an employee of the agency. Both also own some of the stock of the three second-line corporations that own the *Argo Merchant.* Nicholas Skarvelis has more authority than his brother and with Mark Madias runs the affairs of the agency and the ships under its control.

The Armco Financial Corporation, A. G., a Swiss Bank, holds the mortgage for Thebes Shipping. Under the financial and management arrangements for the *Argo Merchant,* all income from its operations went directly to the bank through the Argo Cash Collateral Account. The bank then released money to Amership which paid for the crews' salaries, ship supplies and equipment, maintenance, and any other operational expenses. Any profits that were left over after mortgage payments, interest, and day-to-day expenses were to pass not to Thebes, but directly to the back-up corporations and the stockholders. Thus, Thebes Shipping, Inc., had no income, no expenses, no payroll —not even a checkbook. Its only office was within the Amership offices in New York where Madias worked. Nothing on a doorway in Liberia or New York said "Thebes Shipping, Inc." There was no such name in a telephone book. Other than its listing in *Lloyd's Register of Shipping* as the owner of the *Argo Merchant*, the corporation was virtually invisible. Its only asset was the tanker itself.

When Madias and Skarvelis bought the tanker late in 1973, it already had fourteen reported shipping casualties in ten years. According to statistics kept by McKenzie, who has nineteen thousand tanker accidents on file from 1964 through 1976, the average tanker sustains one reported casualty every four years. Thus, the *Argo Merchant's* accident record was five and one half times worse than average. The vessel was also following a trend similar to other aging, second-hand tankers, his statistics showed. He had determined that ships which change ownership frequently usually have poor casualty records, and that serious casualties often occur shortly after a tanker has been sold. After the vessel was sold to Bootes, S. A., in 1966, it collided with another ship and ran aground in April 1967. When Bootes sold it in 1968, it promptly ran aground in Indonesia. In 1970 it was sold again. In March, 1971, it ran aground off Cape Bruzzano, Italy.

Her new owner in 1973, Thebes Shipping, Inc., had no better luck. The *Argo Merchant* did not run aground immediately, but within weeks of the purchase, on November 16, 1973, her main engine broke down in the Caribbean while she was loaded with crude oil, bound for Puerto Cortes, Honduras. A lubricating oil pump for the turbine had failed. The alarm that was supposed to sound to warn the engine room crew of such a failure also had failed. The ship was towed to Curaçao, Netherlands Antilles, on November 24. All of her turbine bearings were severely damaged. Other engine parts were found in disrepair or were damaged during the installation of new parts. It was March 20, 1974, four months after the breakdown, before the vessel sailed again.

When Thebes purchased the *Argo Merchant*, Madias and Skarvelis immediately arranged a three-year charter for her to

Texaco, which hired the tanker to deliver products to and from its refineries and storage terminals. The four-month lay-up for repairs did not get the arrangement off to a good start, for either Thebes or Texaco. Six months later, on October 18, 1974, her boilers broke down again while she was en route from Deepwater, New Jersey to Baltimore, Maryland, loaded with oil. A steam valve and feed pump had failed, providing too little water to the boilers. Both had burned out and had to be completely overhauled. Repair work took seven months. Thus, in the first year and a half of her charter with Texaco, the tanker was out of service for about eleven months.

On May 14, 1975, after the repairs were completed on the boilers, a Liberian government inspector boarded the vessel for its annual check. He spent four hours on board to inspect everything but the hull and the engine room. He checked items such as extinguishers and lifeboats himself. To check out the navigation equipment—including the gyrocompass, the depth finder, the radar, and the radio direction finder—on the twenty-two-year-old vessel which had not been at sea for seven months, he just asked the master whether those instruments worked. The master told him they did. The inspector also checked the vessel's certificates and the officers' licenses. The tanker passed the inspection. It was required under Liberian regulations to have another one before August 14, 1976.

At the end of July, 1975, the tanker was at drydock for a hull inspection and survey by Bureau Veritas, one of a handful of international vessel inspection societies set up by the shipping industry to regulate the shipping industry. After the survey, Bureau Veritas gave the *Argo Merchant* its highest confidence rating.

Nevertheless, her troubles continued. On February 27, 1976, the *Argo Merchant* was delayed for eleven days in Puerto Cortés because her port boiler fan, which blows air into the boiler for proper combustion of the ship's fuel, had burned out. She sailed again on March 11. It was her last reported casualty until the morning of December 15, 1976, when she ran aground on Nantucket Shoals.

During the three years the company owned the ship,

Madias says, Thebes Shipping spent about two million dollars for repairs, principally to overhaul the turbines and the boilers.

But these major incidents were not the only problems that plagued the *Argo Merchant* during her service under Thebes Shipping, Inc. Minor mechanical problems and minor oil spills had become part of her routine.

On June 26, 1974, she spilled forty gallons of Venezuelan crude oil into Portland, Maine, harbor while discharging her cargo into the Portland Pipeline, which transports oil to Canada. The cargo had leaked through a faulty valve in the pump room. The Coast Guard fined the owners $250. The spill was contained within a boom routinely placed around the vessel by the pipeline company during discharge operations. It was cleaned up immediately.

On August 26, 1975, within a month of the drydock and hull inspection by Bureau Veritas, the Coast Guard, acting on a complaint from Texaco, found that four to five gallons of oil had leaked into the Delaware River from the tanker while she was discharging at a Texaco facility in Westville, New Jersey. The source of the spill was a defective pump room valve that apparently leaked during pumping operations but not when the pumps were shut down. The leak was temporarily patched. Coast Guard officials in Philadelphia permitted the ship to sail, but they notified the Coast Guard Marine Safety Office in Boston of the incident. Boston was the *Argo Merchant*'s next port, and she sailed from Philadelphia that night.

Marine inspection and pollution investigators were waiting for the tanker the next day as it steamed into Broad Sound, the entrance to Boston Harbor, and they went aboard. The inspectors saw oil rising to the surface on the starboard side of the tanker at a rate of four to five gallons per hour. They suspected a crack in the hull or a leaking rivet around the number four center tank was the source of the pollution. After considering turning the ship away, the Coast Guard permitted it to proceed to discharge her cargo at Castle Island, South Boston. An oil spill boom was placed around the ship. But by the time she arrived at the discharge terminal, she was no longer leaking oil. Inspectors estimated the total spill at sixty-five gallons, for which the owners were fined $350.

The incidents at Philadelphia and Boston angered Madias and Skarvelis. The tanker, after all, had just come out of drydock and a hull inspection which had turned up no deficiencies. They maintained the oil in Philadelphia had drifted alongside the *Argo Merchant* after leaking from another tanker. And they insisted the oil in Boston was also from another ship. The reports amounted to little more than harassment from Texaco, Skarvelis later charged in court testimony. Texaco had signed the three-year charter agreement just before the Arab oil embargo in late 1973, and tanker rates were high. Immediately after the embargo, however, the demand for oil dropped around the world and hundreds of tankers were laid up, without cargoes to carry. Transportation rates plummeted. If Texaco could have found reason to break the charter in 1975, Madias maintains, it could have hired another tanker at a savings of $100,000 a month.

But Texaco had other reasons for reporting oil spills that occurred at its facilities. Anyone who spills oil and fails to report it immediately to the Coast Guard is subject to a fine of $10,000.

Thebes Shipping, Inc., appealed the oil spill fines in Boston and Philadelphia, but lost. Texaco did not cancel the charter agreement.

The Coast Guard never actually determined the source of the leak in Boston. But they did find water leaking slowly into the number two port, center, and starboard tanks, an indication, officials felt, of a possible crack in the hull. Rear Admiral Stewart, the Commander of the First Coast Guard District, informed the tanker owners that the *Argo Merchant* would not be allowed back into any first district port—from Providence, Rhode Island, to Portland, Maine—unless she were coming in for repairs in a "gas free" condition—with no potentially explosive gases in her tanks—or until the owners furnished proof that the number two tanks had been repaired.

He also told the owners, "I have recommended to the Commandant of the Coast Guard that entry of the *Argo Merchant* into any port within the United States be denied until these conditions are met."

While Admiral Owen W. Siler, the commandant, agreed in part with Stewart's recommendation, his legal staff could find no authority under which a foreign vessel could be banned outright from American ports unless the vessel lacked a certificate of financial responsibility for oil it may spill. Otherwise, the staff determined, the Coast Guard had to let a tanker come into U. S. waters and board it to determine if it was leaking. If it was, then the Coast Guard could turn the ship away. But if the same tanker approached the same port a week later, the Coast Guard would have to board it once again in U. S. waters before they could refuse it entry into port.

Thus the *Argo Merchant* was not banished from U. S. waters, and on December 19, 1975, she visited Port Arthur, Texas, where she spilled ten gallons of oil. In June, 1976, she stopped at Mobile, Alabama, where the radio direction finder was repaired, and where Georgios Papadopoulos boarded her to take over command.

In July, her main gyrocompass, one of her radar sets, and two radio receivers malfunctioned and were repaired at a Caribbean port. On July 20, while at sea, her crew had to shut down one of the two boilers for minor repairs, slowing the vessel to half speed.

In mid-September, 1976, the *Argo Merchant* made her final stop in an American port. That was New York, where, in contrast to her past chronic minor oil spill problems, she passed a spot Coast Guard inspection that was limited strictly to her cargo discharge procedures. She was not leaking any oil.

The Texaco charter expired in mid-September, and the tanker steamed about the Caribbean on a spot charter arrangement, picking up cargoes on orders from Amership much the way a radio-dispatched taxicab picks up fares. On October 15, while at sea, her engines were stopped for a short time while her crew made some electrical repairs. On October 20 and 21, her port boiler was shut down for adjustments, and on November 11 both boilers were shut down and the vessel was adrift for several hours while more repairs were made. Her owners and her crew described the frequent shutdowns as routine for a tanker, especially one of the age of the *Argo Merchant*, and said they were not an indication of a ship in poor condition. Meanwhile, on November 7, the tanker spilled a small quantity of oil in a port in Ecuador, for which Amership was billed $675 to pay for cleanup costs.

On November 30, Captain Papadopoulos got orders from Amership to sail to Puerto La Cruz, Venezuela, to pick up a cargo of oil bound for Boston. As the ship moved into port on December 2, H. C. Lombard, the Boston agent for the *Argo Merchant*, notified Captain Lynn Hein's Marine Safety Office in Boston that the tanker was due to return there within two weeks. Hein's office immediately notified the Commandant in Washington and Rear Admiral Stewart at First District headquarters a few blocks away. Hein's files showed no indication that the problems inspectors had found on the tanker in August, 1975, had been repaired. So, even before the tanker began taking on her cargo in Venezuela, Hein's office made plans to inspect the ship before permitting her to enter the port.

"Upon entering U. S. waters," Hein's office said in a message to Washington on December 2, "vessel to be boarded by a marine safety officer to verify status of vessel and if found in leaking condition to be ordered to depart U. S. waters."

As she waited for a berth in Puerto La Cruz to take on her cargo, the *Argo Merchant* was more than four months overdue for her Liberian inspection. All of her required certificates, however, were in order. Her officers were properly licensed, and her Bureau Veritas "highest confidence" rating was up to date. But the U. S. Coast Guard considered her a suspect ship, and it was awaiting her arrival in Boston.

16

The Last Voyage

At 6:20 A.M. December 3, 1976, the *Argo Merchant*, with the aid of harbor tugboats, eased into a berth in Puerto La Cruz to receive her cargo. The oil that would fill her tanks was in two large storage tanks owned by the S. A. Meneven subsidiary of the Venezuelan Petroleum Company. The United States gets about twenty percent of all its petroleum imports from Venezuela, and the majority of it goes to ports along the East Coast. Just what tanker carries the oil to these ports has little to do with the Venezuelan Petroleum Corporation or the company that orders the oil. Tankers and their cargoes are matched up through a variety of brokers and middlemen who operate a kind of blind dating service. The goods themselves are what the customer orders, of course, but the package they come in is not necessarily guaranteed.

Exactly who owned the cargo as the *Argo Merchant* pumped her ballast out of her tanks that morning to make room for the oil is not known. The Cirillo Brothers Sales Corporation, also known as Cibro Sales, of New York, an oil broker, had already arranged the purchase of the cargo from the Venezuelan government and on November 18, 1976, agreed to sell it to the Holborn Oil Company, Limited, of Hamilton, Bermuda, a unit of Coast States Gas Corporation of Houston, Texas. Holborn, in turn, agreed to sell it to Northeast Petroleum Industries, of Chelsea, Massachusetts, which owns the terminal in Salem where the cargo was headed.

The tanker completed her deballasting at 4:20 P.M. that afternoon, and the hoses that would carry the oil to her tanks were connected at 5:10. The pumps were turned on ten minutes later, and they churned for twenty-four hours and ten minutes before they were shut down, having completed the task of moving 7,677,684 gallons of Number Six oil into the vessel. It was six more hours before she was loaded with fuel for her own engines, and at 2 A.M. December 5, she pulled away from her berth, out into the harbor. Two hours later, she was "full away," under her own power, having disembarked the pilot that had guided her out of the harbor and shed the tugs that had pushed her.

At 8 P.M. on the first night at sea, Anastasios Nisiotis, the ship's junior officer, came to the bridge to relieve Chief Officer Georgios Ypsilantis, who was completing the four-to-eight watch. Nisiotis, a licensed second officer serving in the third mate's position, glanced toward the helm and noticed that the two men who had been with him on his morning eight-to-twelve watch were back again.

"Am I going to have my watch with these men?" he asked Ypsilantis, his voice showing his annoyance. "I don't want them. Get me some ABs so I can stand my watch. I won't do it with these men. They're just deck boys."

It was customary for two "able-bodied seamen," commonly known as ABs, to stand watch with each deck officer to man the helm and the lookout. In Puerto La Cruz, however, two ABs who had been aboard more than a year took their pay and left the ship. No other crew were immediately available to replace them, so Ypsilantis had to juggle the watch assignments with what he had. He had assigned to Nisiotis' watch Joseph Roach of Trinidad, a deck hand who had done some steering before, and Jose Rivera of Honduras, who had mustered on the vessel as a wiper in the engine room, the lowest paying job on the ship. Rivera had never taken the wheel of a large tanker.

Ypsilantis wanted to avoid a dispute with his fellow officer.

He told him to calm down and take Roach and Rivera for one more watch. They would settle the problem the next day.

Nisiotis, a black-bearded sailor with intense, fiery eyes, went into a rage and Ypsilantis called the captain to the bridge. Papadopoulos didn't bother hearing Nisiotis out.

"If you don't like it, go to your cabin," the captain said. "I'll take over your watch."

Nisiotis stalked over to the ship's log, grabbed a pencil, and in a furious scrawl in Greek wrote:

"At this time, Second Mate Nisiotis, Anastasios, requested from the chief officer to change helmsmen because they are not seamen. One is a deck boy and because of the need, became a seaman having complete ignorance of steering, and the other, a wiper from the engine room, completely unqualified for a seaman's rating. The Captain, having come with the Chief Mate to the bridge, said the following:

" 'If you like them, fine. If not you can go to your cabin.'

"After this, the writer of this entry, Second Officer Nisiotis, Anastasios went to his cabin."

Actually, he never got there. Ypsilantis called the two seamen on his own watch back to the bridge for another four hours and assigned them to Nisiotis for the rest of the voyage. Roach and Rivera would stand the four-to-eight watch with him. Satisfied, Nisiotis remained on the bridge and stood his watch. When he calmed down from the incident, he decided he had erred in recording it in the log. Against ship regulations, he erased it.*

For the next several days, the voyage proceeded smoothly, though a bit slower than Papadopoulos had expected. He had charted a course northward from Venezuela through the Mona Passage, between Puerto Rico and the Dominican Republic, then northwesterly toward Cape Hatteras, North Carolina. Once he reached Hatteras, he planned to turn northeasterly on a course for the Nantucket Lightship before rounding Cape Cod and heading into Boston Harbor. The waters of the Carib-

---

*The text of the entry was recovered through the efforts of a document specialist and a translator during court proceedings after the grounding.

bean, south of Mona Passage, were calm, as were those of the southern North Atlantic on the other side of the island nations. The ship sailed much of the time on automatic pilot. Two days out of Puerto La Cruz, Papadopoulos sent a message to the Amership agency, saying he expected to arrive in Boston on the afternoon of December 13. The next day, however, he revised that to 8 A.M. the fourteenth, and by December 10, five days into the voyage, the captain had pushed his arrival time back even further, to noon, December 15.

Papadopoulos was under instructions to make sure the temperature of his cargo was at least 120 degrees when he arrived in port so it would pump easily, and that may have contributed to the slower progress. The same steam that powers the ship heats the cargo tank heating coils, and thus any steam used to keep the oil warm steals from the speed of the ship. On December 7, the average cargo temperature in the thirty tanks was 119 degrees. By December 10, it was down to 108.2 degrees. It is not unusual for a master to purposely let the cargo temperature drop early in a voyage to make better time, then rush to get it back up again just before arrival in port. But Papadopoulos couldn't afford to use such tactics, at least not for long. Some of the heating coils were broken. On December 11, when the average temperature dipped to 107.7 degrees, he reassured Amership that his crew had set up portable heaters in an effort to increase it again. But he didn't have enough steam hose on board to rig all the tanks that needed heat, and he asked his agent to have 1,500 feet of it waiting for him when the tanker arrived in Salem. The portable units did help. On December 12, the temperature was up to 112.4. But once again, he adjusted his expected arrival time; he figured to be in Salem by 5 P.M. the fifteenth.

At 11 P.M. December 12, Nisiotis was on watch working with the radio direction finder, the RDF, and picked up the signal of the Diamond Shoals Light station off Cape Hatteras. The station, like the Nantucket Lightship, is a major navigation marker, and it warns mariners of dangerous waters off the North Carolina coast. The radio signal was directly off the bow of the *Argo Merchant*, indicating the light station was dead ahead. When the vessel was within twelve miles of the light, Papadopoulos, who had joined Nisiotis, ordered a change from the northwesterly course the ship was on to a northeasterly heading already laid out on the ship's chart. As the tanker turned to starboard, Nisiotis spotted the Diamond Shoals' beacon in the distance, shining against the black sky. He did not take another bearing with the RDF.

The new course was 040, and it led to a point three miles southeast of the Nantucket Lightship. Mariners use the 360 degrees of the compass in charting and calculating their paths over water. North is zero or three hundred sixty degrees and the numbers increase clockwise around the compass, through the eastern, southern, and western quadrants before coming back to north. Zero-four-zero, as the *Argo Merchant*'s new course would be read, is almost northeast. A course of one-eight-zero, 180 degrees, for example, would be due south, while three-one-five would be northwest. The course was based on a "true north" compass reading from the tanker's electronic gyrocompass, which the captain and his officers relied on almost exclusively to follow their main course. A gyrocompass resembles a giant electrically driven top which operates in an enclosed room below the bridge. It aligns itself precisely with the earth's true North and South Poles, and for that reason gives what are called "true" readings. It is considered more accurate than a magnetic compass, which is affected by the earth's magnetic fields and points to magnetic poles. The difference between a true north and a magnetic north compass reading is known as variation and it differs from place to place on the earth's surface.

The *Argo Merchant* also had a magnetic compass, which the officers and helmsmen could use to verify their gyro course by applying whatever variation is required from their position at sea. Off Cape Hatteras, the variation is eight degrees west; that is, magnetic north is eight degrees west of true north. As the ship moved up the coast toward Cape Cod, the westerly variation would increase.

During the turn off Cape Hatteras, Nisiotis watched the

gyrocompass repeater, an instrument to the right of the ship's wheel that included a disc numbered from zero to three-sixty that rotated to give the gyrocompass reading. The repeater made a click, click, click sound as the bow swung through each degree of the course change. It finally settled at 040. Satisfied, Nisiotis looked at the magnetic compass to the left of the wheel. Its needle pointed to fifty-four degrees. He walked over to a blackboard within easy view of the wheel, erased three numbers that were posted there, and entered three new ones: 040 true, 040 gyro and 054 magnetic.

The variation between the compasses was fourteen degrees, six more than were indicated on the nautical chart off Cape Hatteras. Magnetic compasses are also affected by metals aboard the ship, a factor known as deviation, which is checked by tests and noted on a card near the compass. On a northeasterly course, the tanker's magnetic compass had no deviation. Thus, there was an unusual difference between the compasses of six degrees. Papadopoulos and Nisiotis were sailing by gyro, however, and other than perfunctorily recording the magnetic reading in the ship's log and on the blackboard, they gave the difference no more than passing notice. Papadopoulos returned to his cabin. Georgios Dedrinos relieved Nisiotis at midnight.

Four hours later, at 4 A.M. on the thirteenth, at the end of his watch, Dedrinos conferred briefly with Ypsilantis, who was to replace him. Both had noticed the wind, which had been blowing from the west at fifteen knots, had freshened to nearly twenty-five knots and was now coming from the northwest. That meant the seas were surging against the port side, having the effect of pushing the vessel to the east of its course, away from shore. Dedrinos awoke the captain with a telephone call, since Papadopoulos was the only one on board who could authorize a course change, and advised him of the weather.

"Okay," Papadopoulos said. "Change the course four degrees to port, to zero-three-six."

Dedrinos ordered the helmsmen to turn the vessel slightly to the left, toward shore. When the gyro stopped at 036, Dedrinos checked the magnetic, then went to the board and wrote three new numbers: 036 true, 036 gyro, 050 magnetic. The fourteen-degree variation remained the same. Dedrinos entered the figures in the log and went to his cabin.

At noon each day, the officer on watch, usually Nisiotis, took a solar fix to determine the ship's exact position in latitude and longitude. It was the only time each day that the officers plotted their position on the chart, unless special circumstances such as an approach to a light or a port required additional fixes, and it served as a daily bench mark to which the previous day's progress was measured and from which the next day's course was projected. At noon on the thirteenth, Nisiotis, using a sextant, measured the angle of the sun above the horizon, and with earlier measurements taken of stars, used tables to convert the numbers to a position on the chart. His calculations seemed to prove that Dedrinos was right in figuring the wind was pushing the vessel away from shore. The tanker was ten miles east of the 040 course line. It was also about 145 miles east of Norfolk, Virginia, and further enough north to be in an area where the variation had increased to ten degrees west. The difference on the compasses remained fourteen. They didn't take another fix until noon the next day.

Sometime during the midnight to 4 A.M. watch on the fourteenth, the magnetic compass reading changed three degrees to the west, to 047. The gyro, however, remained steady at 036. Dedrinos, who was on watch at the time, noted the difference in the log, but was not concerned that one compass had changed while the other had not. He assumed that the course had remained steady at 036.

At noon on the fourteenth, Nisiotis took another solar fix. He plotted the vessel's position about 175 miles east of the Delaware coast. It was a particularly important fix because it would be the last one before the vessel entered the waters marked by the Nantucket Lightship. So, at 2 P.M., the captain asked Dedrinos to take another fix to verify the first. He did, and after adjusting the position for two hours of sailing, he verified Nisiotis' noon position. It was four miles west—toward shore— of their projected 040 course. And, measured back from noon the previous day, their projected track pointed inside the lightship, toward the shallow sandbars of Nantucket Shoals.

Papadopoulos, however, decided against a course change then. He wanted to get a 6 P.M. position with a star sighting before deciding whether to change his heading.

Meanwhile, in Boston, Captain Hein's office got word from Lombard, the tanker's Boston agent, that the *Argo Merchant* was due in Salem about 3 P.M. the next day and would be off Provincetown at the tip of Cape Cod about noon. The Coast Guard Air Station at Cape Cod was alerted and asked to make a surveillance flight over the vessel off Provincetown to see if she was leaking oil. The pilot was to file an immediate report to Boston. The Coast Guard also notified Ralph Hobbs, Jr., the harbormaster in Salem, of the vessel's pending arrival and, at one point on the fourteenth, called to make sure that the tanker had not already reached its port.

Meanwhile, at 1 P.M. Papadopoulos sent another message to Amership. The cargo temperature was down to 104.8 degrees, the lowest it had been on the entire voyage. And, having been slowed by rough weather during the past two days, he informed the agent that he now expected to arrive in Salem at 9 P.M. on the fifteenth.

Within eighteen hours, however, the captain's daily cargo temperature reports and his almost daily revisions in his arrival plans would be meaningless.

Shortly before 6 P.M. December 14, Ypsilantis noticed that the gyrocompass repeater next to the wheel was giving erratic readings that wavered six to seven degrees on each side of the 036 reading. He notified the captain. After Papadopoulos inspected the compass, he ordered the helmsman to steer by the magnetic compass. The gyro was broken. It was an old compass and its instruction manual said constant shipboard maintenance was important to its reliable operation. No one on board, however, knew how to maintain or repair it.

Exactly what happened to the gyrocompass and the repeater is open to question. It is possible it had been malfunctioning for at least fourteen hours before Ypsilantis noticed the sharp jumps in the repeater, for it was about fourteen hours earlier that the magnetic compass shifted to 047 while the gyro held steady at 036. And the unusually wide variation between the compasses during the latter part of the boyage suggests that the compass may have broken even earlier. The captain and his officers didn't think so. They thought the magnetic compass had been affected by large mineral deposits on the ocean bottom or ashore. It was an explanation right out of a navigation textbook, and it was improbable. The route between Cape Hatteras and Cape Cod is heavily traveled and well charted. Any such deposits that would throw a magnetic compass off as much as six degrees would be clearly marked on a nautical chart. Whatever the explanation for the differences, Ypsilantis and Papadopoulos did not consider at 6 P.M. that night when the gyro was clearly broken, nor at any time during the next twelve hours, that the faulty compass might have led them off their course.

It was cloudy at 6 P.M., and the officers were unable to take the star sightings Papadopoulos had planned on before deciding whether he needed a course change. The men couldn't see any stars. So the captain stayed with his course of 036 true, steering 047 on the magnetic compass. By then, however, the vessel was far enough north that the normal variation between magnetic and true was fourteen degrees. Thus, it is possible that the vessel may have been making a 033 course true, three degrees closer to shore than Papadopoulos had thought.

Shortly before midnight, Nisiotis noticed that currents seemed to be pushing the vessel slightly off course, toward shore. He remembered that his noon fix had already showed the vessel was to the west of the course line, so he asked the captain about turning slightly to the east, to starboard. Papadopoulos agreed to a three-degree change. The helmsman turned the wheel until the magnetic compass read 050.

At midnight, Dedrinos came to relieve Nisiotis, and together they plotted a "dead reckoning"—their best guess of where the ship was—by running a line up from the 2 P.M. fix along the 040 course line for about eighty-five miles—ten hours at eight and a half knots. The position was thirty-six miles southeast of the Nantucket Lightship. Dedrinos expected he would see the beacon from the bridge by 3 A.M. Just before 1 A.M., he called the captain to the bridge, and the search for the lightship was on.

During the first two hours of the search, Papadopoulos and Dedrinos used a November current chart, which indicated that west to east currents in the areas south of Nantucket would push them seaward of their course. To compensate, Papadopooulos held fast to a course that pointed his ship slightly to the shore side of his course. If they had had a December current chart on board, they would have noted that currents move east to west in the Nantucket area during that month, and were, in fact, pushing them toward shore. Their course was compounding the error, not correcting it.

When Ypsilantis came to the bridge for his 4 A.M. watch, he came with the two inexperienced helmsmen he had been forced to use because the ship was short-handed and because Nisiotis had refused to work with them. Jose Rivera, the least experienced of the two, was at the wheel from 4 to 5 A.M. Joseph Roach was on the lookout. They switched positions at 5, and they were to change again at 6. After 5:45, just before he was to take the helm, Rivera went below for coffee. When she ship ran aground fifteen minutes later, he was not on the lookout.

When Papadopoulos and Ypsilantis worked with the RDF to find the signal of the Nantucket Lightship, they were using an old instrument which had not been calibrated in nineteen months, in violation of international shipping treaties that require such instruments to be checked for accuracy at least once a year. When Nisiotis had used it off Cape Hatteras two nights earlier, he picked up the signal of the Diamond Shoals Light directly off the tanker's bow. Then he saw the beacon ahead of him as the ship made a turn. When Ypsilantis finally picked up the Nantucket Lightship's radio signal at 4:30 A.M., the signal was, again, directly off the bow. In fact, however, the *Argo Merchant* had passed to the port side of the lightship, and at 4:30 A.M. it was well to the vessel's starboard side, about 120 degrees off the bow.

At 5 A.M. and 5:30 A.M., Papadopoulos noticed readings of fifteen to twenty fathoms—ninety to one hundred twenty feet

—on the ship's depth finder. The tanker's draft was thirty-five feet. Despite depths of more than two hundred feet along his intended 040 course line, and despite indications on the chart that such shallow depths could mean only that he was heading for even shallower water, the captain ignored the depth finder, relying instead on the signal from the RDF. He also ignored an explicit warning in the Atlantic Coast Pilot, a detailed navigation manual for U. S. waters which he had on board. Of the Nantucket Shoals, the manual says:

> Approaching this section of the coast is dangerous for all vessels because of the off-lying banks and shoals, the strong and variable currents, frequency of fog and the broken nature of the bottom. Soundings alone are of little value in establishing the position of a vessel, but the depth should be checked frequently to insure that the vessel clears all dangers. In thick weather especially, the greatest caution is necessary, and vessels equipped to do so should make good and timely use of the electronic aids to navigation systems to check their position frequently. The depth should never be shoaled to less than fifteen fathoms without an accurate fix having been obtained, and it is advisable to remain offshore in depths of twenty fathoms or more.

But Papadopoulos had no loran set on his ship "to make good and timely use of" in an effort to determine his position. Ypsilantis was acutely aware of that as he struggled to take a 5:30 A.M. fix when it was too dark to see the horizon. When he added wrong, the position he came up with was more than fifty miles off. Even if he had done his arithmetic correctly, however, he would have had a twenty-mile error. With a loran set, he could easily have determined the ship's position accurately to within a length of the vessel itself.

The owners of the *Argo Merchant* had paid $2.4 million for the ship and had apparently invested another $2 million to repair it after several serious breakdowns. When fully loaded, the cargo it carried was worth more than two million dollars. Yet, Thebes Shipping, Inc., had seen no reason to spend $5,000 to $10,000 to equip the tanker with loran.

Thus Papadopoulos and his officers were left to navigate the vessel by the stars—impossible when it is cloudy or dark—

with outdated charts, with a broken gyrocompass, and with an RDF that hadn't been checked for accuracy in more than nineteen months.

But, as he stood on the bridge at 5:45 A.M. on December 15, 1976, peering out over the waters in search of the lightship, Papadopoulos knew he hadn't had an accurate fix for fifteen hours; he knew his gyrocompass had not worked properly for at least the past twelve hours; he knew Ypsilantis' 5:30 A.M. star fix indicating they were well to the southwest of the lightship was improbable; he knew the RDF calibration was out of date; he knew he was sailing over ninety feet of water; and he knew he should have seen the Nantucket Lightship two hours before. He didn't know where he was.

Nevertheless, he did not slow down. He did not change course. He did not stop.

When the *Argo Merchant* ran hard aground at 6 A.M., she was in eighteen feet of water thirty miles north of the Nantucket Lightship and fifteen miles from the nearest shipping lane. She was also twenty-four and one half miles off her intended course of 040.

In October, 1977, after thousands of pages of testimony from the captain, his officers, and crew were compiled and sifted by lawyers and maritime experts, the lawyers for Thebes Shipping, Inc., seeking to limit their client's liability in the accident, maintained in U. S. District Court in Manhattan that the *Argo Merchant* went aground because of human error. Papadopoulos, they said, had set the wrong course for Nantucket Lightship from his last known position at noon December 14; had not allowed for variation between the magnetic and true compass readings; had not used his depth finder soon enough though he was approaching shallow water; and had not properly used the RDF in trying to get a bearing off the Nantucket Lightship.

About the same time, the Liberian government, which had investigated the accident on its own, also decided human error led to the accident. It revoked Papadopoulos' master's license, suspended Ypsilantis' first mate's license for four years, and lifted Dedrinos' second mate's license for nine months.

Lawyers for the Continental Insurance Company, seeking to recover the $2,221,958.31 they paid to the owners of the cargo, maintained that the tanker went aground because it was unseaworthy. The *Argo Merchant* was not seaworthy, they said, because it was not equipped with a December current chart or an up-to-date navigation chart of the Nantucket area; because her gyrocompass had not been properly maintained and was inoperative just before the grounding; because the RDF, which also had not been properly maintained and calibrated, was broken just before the stranding; and because the *Argo Merchant* was not equipped with loran.

It would be up to a federal judge and, almost certainly, an appeals court, to decide in the end why the *Argo Merchant* ran aground.

Just who actually owned the oil when it began leaking from the tanker's hull was never determined. The Venezuelan Petroleum Association billed Cibro Sales for $2,019,962.10 for the cargo, or $11.05 per barrel. Cibro then billed Holborn Oil for $2,184,493.90, or $11.95 per barrel. The cargo was to be sold to Northeast Petroleum for $2,221,958.31, or $12.155 per barrel. Although these agreements had already been worked out as the ship sailed toward Salem, when the tanker ran aground, each of the firms said one of the others owned the oil. The insurance company paid the claim under a legal agreement in which none of the companies had to admit ownership.

The *Argo Merchant* struck the shoal six hours before a Coast Guard crew was to take off from the Cape Cod Air Station to fly over her and check for oil leaks, and nine hours before a Coast Guard inspection team was planning to board her in Boston Harbor. Thus, their concern about the tanker's condition had gone for naught. A week later, after the ship had broken apart, the Coast Guard and the other federal agencies which had gathered on Cape Cod in response to the accident no longer had a ship to worry about. Instead, they had more than seven and one half million gallons of oil spreading over the ocean. It was the largest costal oil spill in U. S. history.

17

The Slick

The first dead birds showed up on the beaches of Nantucket on December 21, the day the tanker cracked in half. Their oiled wings and breasts gave islanders tangible evidence of what they had seen previously only in newspapers and on television. Birds are among the first and most visible victims of almost any coastal oil spill, and their oil-gummed feathers and helpless, bewildered expressions have become a pathetic symbol of the ravages of oil against nature. They are especially vulnerable because they seem to be attracted to oil. A slick absorbs the impact of waves and appears to flatten the sea, providing a deceptively inviting resting place for birds. Diving birds, such as loons, murres, and auks, which are prevalent off Cape Cod, become virtually covered with oil, or, as it is sometimes called, oil-logged, when they dive into a slick or surface into the gunk after a dive into cleaner water. Once covered, anything a bird does to try to cleanse itself works against it. The oil breaks down the insulating quality of their feathers, making birds susceptible to pneumonia or freezing to death. It also affects their buoyancy and some drown. Some become so preoccupied with preening to get the oil off them that they forget to feed, and finally become so exhausted they can't feed and they starve to death. Or they ingest the toxic oil, which poisons them.

The only effective solvents for cleaning the birds are petroleum-based, and thus have the same effect of exposing them to the cold by breaking down their natural oils. After treatment, they must be kept inside and fed until the oils are replenished, a process that takes weeks or even months. It is an expensive and usually futile task. Few birds survive it. After the *Torrey Canyon* spill in 1967, a study of the bird-cleaning operations concluded: "Despite the enormous expenditure of time and effort and large sums of money, present attempts at rehabilitation of oiled seabirds are frankly worthless. If anything, present evidence suggests that the rehabilitation of oiled seabirds satisfies a human need, but has no biological significance whatsoever." Little has been accomplished since to alter that assessment.

Nevertheless, about eighty Nantucket residents turned out at a special meeting on the night of the twenty-first to volunteer to patrol beaches and help rescue and clean birds that came ashore alive. It was not only a human need that drew them to the meeting. They felt helpless against the slick itself and just needed to take some action, to react constructively in the face of its threat. The wildlife victims were their outlet. Jim Lentowski, executive director of the Nantucket Conservation Foundation, a nonprofit organization, devised a plan in which the island's shoreline was divided into five sectors and a coordinator was appointed for each one. A bird collection center was set up at Nantucket Airport to serve as a refuge for live birds until they could be flown for cleaning to the Felix Neck Wildlife Sanctuary twelve miles away on Martha's Vineyard. The dead ones would be flown to Cape Cod for freezing and eventual autopsy.

But because birds are considered the property of the federal government and no one else could authorize anyone to pick up or clean up an oiled bird, few were actually removed from the beaches until representatives of the U. S. Fish and Wildlife Service arrived at noon the following day. Thus some of the birds were left on the beach for at least a day and a half. The delay did little to calm the anger and frustration of the islanders, who had expected a quicker response from federal officials.

When they did arrive, members of the Fish and Wildlife

Service surveyed the island and counted 125 oiled gulls among three thousand that they saw. The following day, the twenty-third, a state and a federal biologist surveyed the water near the wreck in an airplane and spotted 31,000 birds, most of them between Nantucket and the tanker or between Nantucket and Martha's Vineyard. Both locations were out of the huge tract of sea to the east of the *Argo Merchant,* where the oil was concentrated. Thus the biologists found no immediate need for the noise and whistle bombs they had brought to the island to frighten birds away from the slick. They concentrated instead on the beach patrols and the cleaning. By noon, December 24, twenty-eight oiled birds had been taken to the bird collection station. Eighteen were alive; ten were dead.

Oil's impact on fish, shellfish, and other marine organisms is not as readily visible as it is on birds. But scientific studies show clearly that it is as cruel and debilitating to scallops, lobster, fish eggs, and animal plankton, for example, as it is to birds. When the barge *Florida* went aground outside of West Falmouth Harbor, Massachusetts, on September 15, 1969, it spilled nearly 200,000 gallons of Number Two fuel oil. Before the oil even reached the town's rich scallop beds, the scallops became disoriented and sluggish. Clams came up out of the mud with their necks hanging out of their shells as if they were dying of thirst. They didn't. They suffocated. Within three days, the shoreline along the lucrative shellfishing areas of the community were lined with scallops, oysters, clams, crabs, and lobsters, all dead. In some areas, oil had coated the mud flats at low tide when the sea water had dropped out from under the goo. In others, it appeared that no oil had reached the shellfish beds, but toxic compounds apparently had dissolved in the water and poisoned their victims before the black goo itself got to them. Scientists also found fish that normally live on the bottom swimming aimlessly near the surface, as if they had been drugged. Because the spill occurred near the Woods Hole Oceanographic Institute, it was thoroughly studied and documented by scientists there. It is one of the most devastating spills on record. When the *Argo Merchant* ran aground on December 15, 1976, some shellfishing beds in West Falmouth were still closed, seven years and three months after the incident.

Not all spills are as damaging or as dramatic in their impact on marine life. But oil's effects do not end with suffocating or poisoning its victims. Scientists once were preoccupied with organisms that were killed by oil; now they are becoming increasingly concerned with what they call "sublethal" effects of oil pollution, those that don't kill victims outright but that harm them in more subtle ways that may lead to their deaths or cause abnormalities in growth, development, and reproduction. Oil that contaminates fish eggs on the ocean surface can either kill the embryos or inflict genetic damage that will kill them or affect the next generation of fish. Many marine organisms communicate with chemical odors they give off into the water. Even minute quantities of oil, studies have shown, can interfere with these intricate sensing processes enough so that a male lobster is confused responding to signals given off by his prospective mate, that a snail misses chemical odors emitted by its potential prey, or that a salmon refuses to return to its home river for spawning. The sexual organs in mussels contaminated with oil have not developed normally, while cell division in phytoplankton has been inhibited. Respiration in many forms of marine life is reduced after contact with oil. Whales can suck it into their lungs through their blowholes. And organisms such as clams, exposed to chronic concentrations of oil, have developed tumors.

Some of these effects are temporary; some are permanent. All are poorly understood by scientists, and their ultimate impact on each organism as well as the entire marine community is a mystery.

Dr. Eric D. Schneider, director of the Environmental Protection Agency's Marine Water Quality Laboratory in Narragansett, Rhode Island, the agency's leading research facility for oil in the marine environment, is often asked what levels of petroleum hydrocarbons, the toxic compounds of oil, are safe in the ocean.

"That depends," he answers. "What do you want to save?"

The question might also be phrased: "What are we willing

to kill?" Some of the sublethal effects have been detected in studies when the concentration of oil is as small as ten parts per billion—in effect, ten drops of oil for one billion drops of water.

Though the specter of black goo washing up on Nantucket and Cape Cod beaches was a bleak image indeed, scientists watching the *Argo Merchant's* spill were more concerned about salt marshes, shellfishing beds, and other fragile ecological areas, the breeding grounds for hundreds of forms of marine life. These sensitive areas would suffer much longer than an open beach, which would clean itself quickly under the constant pounding of the winter surf, the scientists figured. And while the blanket of oil spreading from the tanker was threatening birds and some fish eggs, the scientists were more concerned that some of the cargo would sink. If it did, it would mix with the bottom sediments, a habitat for plankton, the tiny organisms that make up the base of the marine food chain. In addition, apart from the influence of the wind, the scientists knew that any oil on the surface would probably be carried seaward by the prevailing currents in the area. Oil that sank to the bottom, however, would move slowly toward shore and could carry the tanker's cargo to the coast unnoticed.

As that information was circulated, the Coast Guard was under pressure from news reporters to determine whether the oil was indeed going to the bottom. On December 23, two Navy divers who were put aboard the *Vigilant* to take underwater photographs of the slick and conduct current tests were also asked to check the bottom for traces of oil. Two divers went down in 140 feet of water along the path of the slick. They found clean, white sand, covered with clams. One of them brought a handful of sand back to the *Vigilant*. It was clean. Commander Cruickshank reported the findings to the beach.

That night, Cruickshank watched the evening news on the television aboard the cutter and saw and heard Walter Cronkite declare that the *Argo Merchant*'s oil was not sinking. Cruickshank cringed. He knew the divers' observations were hardly scientific. Droplets of oil that had been churned up in the rough seas might have settled to the bottom and they wouldn't be visible in a handful of sand. He shook his head. Once Walter Cronkite gives a fact like that to the American people, he thought, it's pretty hard to take it back.

Hordes of scientists had already descended upon Cape Cod, however, intending to make a much more scientific evaluation of the *Argo Merchant*'s oil, its path on the water, and its impact on marine life. They had started trickling into the Coast Guard Air Station on the night of the grounding. By the breakup six days later, scientists and technicians from the Coast Guard, the National Oceanic and Atmospheric Administration, the Environmental Protection Agency, the Woods Hole Oceanographic Institute, the University of Rhode Island, the Massachusetts Institute of Technology, the University of Southern California, the Marine Biological Laboratory at Woods Hole, the National Marine Fisheries Survey, the federal Bureau of Land Management, the National Aeronautics and Space Administration, the National Weather Service, the U. S. Navy, the Massachusetts Division of Fish and Game, and the Manomet Bird Observatory were involved in the effort. No one was in charge. Unlike the carefully laid out contingency plans for cleaning up oil spills, plans for assessing their environmental impact did not exist. Researchers began working on individual projects almost immediately. There was little coordination. Despite the abundance of talent that had reported to the spill, within a few days the scientific effort was on the verge of chaos.

By chance, one informal group of scientists did have a plan to study oil spills, and they had some money to execute it. They were among the first researchers to arrive on Cape Cod. Known as the Spilled Oil Research Team, with the obligatory federal acronym SOR, it was composed of scientists from the Coast Guard and the National Oceanic and Atmospheric Administration, or NOAA. The team had been set up informally as a kind of research strike force designed to track and study oil spills for the Bureau of Land Management, the agency in charge of the nation's offshore oil development program along the Outer Continental Shelf. The group was particularly interested in accumulating information to help predict how oil that spilled during offshore drilling operations would move over water, where it would go and what areas would be endangered. Thus, they needed oil spills in the open ocean for their studies. When it was organized in February, 1976, there was not much enthusiasm among the members for purposely spilling oil for research purposes. Instead, they were looking for "spills of opportunity," any spills in open water anywhere along the U. S. coast, and they had a million-dollar budget to spend on their studies.

However, the team's interest was confined to tracking and observing oil in the water. They were not set up to study the impact of the oil on marine life and wildlife. That responsibility belongs by law to the Environmental Protection Agency, but it had no organized unit to conduct such a study and no money budgeted to spend on one.

The SOR team, headed by Dr. James Mattson, set up headquarters in Hyannis, Massachusetts, several miles down the road from the intense activity at the air station. They wanted to keep a low profile and do their own work, which included tracking the oil for Captain Hein, and stay out of everyone else's way. During the first few days, however, Mattson and other members of the team spent several hours in meetings with scientists from the University of Rhode Island and the Woods Hole Oceanographic Institute, in an effort to set up cruises on research vessels maintained by both institutions. At the same time, the team gathered current meters and probes, oil sampling buckets and drift markers. While the Coast Guard Strike Team tried to save the tanker, the SOR team began taking measurements.

Meanwhile, Eric Schneider of the EPA's water quality laboratory wondered why no one had called him about the spill. His office overlooks Narragansett Bay within one hundred miles of the wrecked tanker. But despite his facility's responsibility for research on oil in the oceans, no one called him about the *Argo Merchant* during the first several days of the incident. Finally, on the twenty-first, five days after the grounding, he called the EPA regional office in Boston. Officials there were eager for his help and asked him to organize the environmental assessment program. Schneider called a planning meeting for Wednesday, the twenty-second, and invited scientists from about twenty research agencies, including many that were already working on the spill. One of them was Mattson. But the leader of the SOR team declined to attend the EPA meeting, saying he had too much work to do on his own project.

At the meeting, the scientists prepared a tentative plan for assessing the impact of the spill on marine life. A thorough job, they estimated, would cost three to four million dollars. That kind of money wasn't available anywhere.

Determining just how harmful oil is to the environment is difficult work. The science itself is so young that research scientists are still discovering what the questions are, never mind trying to find the answers. There are few established methods for conducting the research, and in fact most scientific studies on oil in the environment conducted before 1970 are now considered unreliable because methods were imprecise, because laboratory conditions came nowhere near simulating actual ocean conditions, and because experiments were not properly designed. It is also an unusually complicated science, because the ocean is comprised of thousands of chemicals in nature, even before oil, a mixture of many complex chemical compounds, is poured into it, setting off untold numbers of reactions that could affect marine life. And to make matters more difficult, each oil spill differs significantly from others before it. For example, *Petroleum in the Marine Environment,* a 1975 publication of the National Academy of Sciences, notes that the effect of oil on marine life depends on what kind of oil is spilled, where it is spilled, what season it is spilled in, what weather and ocean conditions are prevalent at the spill site, what kinds of marine life live in the area, and what methods are used to clean up the spill. With such a wide variety of variables, it is often misleading to compare a spill such as the *Argo Merchant*'s with a previous spill to predict its impact. Oil spreading over the open ocean near a lucrative fishing area in the winter is very different from oil penetrating a fragile salt marsh in the summer.

In addition, oil weathers and changes rapidly once it hits the water. Particularly harmful oil compounds, such as those found in crude and Number Two heating oils, apparently do their damage quickly and then evaporate or dissipate in other ways, while residue from other less volatile components of oil appear to get more dense and sometimes sink to the bottom, either of their own weight or after picking up particles of debris and other matter that build up in the goo and eventually pull it under.

The *Argo Merchant* was carrying Number Six oil, one of the least volatile petroleum products, since most of the harmful compounds remain with lighter products such as gasoline and light heating oil yielded from crude oil in the refining process. However, the thick oil still contains toxic components. And scientists learned through their research that the tanker's cargo had been thinned with about twenty percent "cutter stock," a more volatile, lighter weight oil that made pumping the product easier.

Schneider read Mattson's absence from the December 22 planning meeting as the first shot of a battle between the NOAA and the EPA over who would coordinate the scientific response to the spill. He was right. A few days later he received vague instructions from Washington about the EPA's role, an indication, he felt, of high level maneuverings in Washington for control at the scene of the incident. The Environmental Protection Agency had assessed the impact of nineteen other spills, but it was not prepared for this one. NOAA already had its armies in position at the spill, and it also had money. Meanwhile, Captain Hein, who was not particularly concerned about who coordinated the scientific effort as long as someone handled the deluge of questions and requests for logistical support scientists were pushing on him, asked Mattson's unit to handle those.

By the end of December, NOAA had assumed command and the EPA was playing only a minor role. Despite all the confusion, four research cruises were completed or underway during the final week of the year, with scientists gathering information on the environmental effects as well as the physical transport of the slick. Nevertheless, it was at least ten days before some individual scientists from some of the research institutes knew whom to report to and where they might seek funding. Some felt the confusion cost them the collection of valuable information immediately after the oil hit the water.

While the tanker was still together, dozens of people from around the country called Captain Hein's office to suggest ways of keeping its cargo out of the sea. One person maintained that liquid hydrogen should be carted out to the ship and used to freeze the oil inside the tanks. Another recommended soaking hundreds of big blankets in diesel fuel, dropping them over the ship from a helicopter, and lighting them on fire. That would create enough heat to ignite the oil in the tanks. These suggestions, and others like them, were politely rejected.

Once the oil was on the water and the best containment equipment on the market was lying useless just off the runway at the air station, undaunted citizen-experts supplied Hein with novel ways to collect oil. Cut the tops off all the trees on Cape Cod, one caller suggested, and use them as brooms to soak up the oil. Hire all the fishing boats in the area and have them haul their nets around in tandem to corral the oil, another recommended, apparently without considering that the slick would flow right through the nets. Ring the tanker with hundreds of barges anchored bow-to-stern in a fifty-mile radius of the grounded vessel, another person suggested. Those, too, were turned down.

Captain Hein and Rear Admiral James P. Stewart, commander of the First Coast Guard District, had a hard time rejecting one suggestion, however, It came from Senator Claiborne Pell, a Rhode Island Democrat, a captain in the Coast Guard reserve, and the author of several pieces of legislation in Congress dealing with the ocean and ocean study. The library at the University of Rhode Island School of Oceanography is named after him. At Senator Kennedy's hearing on the night of December 22, Pell said the effort to save the tanker made him proud to be in the Coast Guard. But he did not hide his displeasure that they had not tried to blow the tanker up while it was still in one piece.

"Why wasn't the thought given of trying another explosive

device, like a torpedo?" Pell asked. "Why wouldn't a torpedo do this? I remember in World War II, when I was at sea, in even colder weather than now and saw ships in the North Atlantic burning in the harbor. We circled the ship a day and a half and saw it burn. Why didn't a torpedo have any value effect?"

Capt. Frederick Schubert of the marine environmental protection division at Coast Guard headquarters in Washington told him that products carried in World War II tankers included aviation gas, gasoline, and diesel fuel, which are much more volatile than Number Six oil and would naturally explode when hit with a torpedo. He also noted that the British tried bombing the *Torrey Canyon* to open up her tanks and ignite her cargo of crude oil. Some of the light elements of the crude burned, but most observers considered the effort a failure. The oil that didn't burn was a thick gooey substance, about the same consistency as Number Six oil.

It was too late by then to blow up the tanker, but Pell wanted to try to burn the oil slick anyway. A Massachusetts chemical manufacturer had told Pell about a wicking agent his firm made that would draw the oil like a candle wick into a flame that could be started with a second chemical, a burning agent. The president of the company, Paul Tulley, was with him at Kennedy's hearing. When Admiral Stewart noted that he had a sample of Number Six oil in a bottle in his brief case, Pell wanted to try an experiment on the spot.

"I am willing to go out in the men's room with it and try it," he said. "In the head. We have got the oil and we have got the stuff. Let's try it."

Kennedy suggested the experiment could be run later in the hearing. Pell raised the issue again when representatives of the Environmental Protection Agency took the witness stand. Kenneth Biglane, the man who helped organize the nation's oil spill response plan nine years before and who was now in charge of the agency's Oil and Hazardous Substances office, told the senator that burning the slick was impossible.

"What you have here, you've got blobs of oil, Number Six, floating . . . right at or below the waterline," Biglane said. "If you would look at this oil from an aerial posture, you might see a patch of oil the size of a garbage can lid, for instance, but under that patch would be perhaps a blob thirty feet in diameter. Now it is difficult for me to understand how one could apply any kind of whiffle dust to create a combustion at the surface of the sea in six to eight foot seas."

But Pell was persistent, and he did get the Coast Guard to agree to try to burn the tanker's slick. He lost his battle for the men's room experiment, however, saying that he learned that unless you have a revolution, "you can't set a fire in Boston without the permission of the fire department."

Besides, added Russel Train, director of the Environmental Protection Agency, such an experiment would clearly violate air quality standards.

On December 24, the necessary chemicals were taken out over the slick by helicopter. Pell went in an airplane to observe the experiment. But the Coast Guard crewmen could not find any blobs of oil big enough to burn. The slick seemed to be dispersing into small patches. The senator returned to the air station without seeing his plan tried.

Meanwhile, the oil slick had continued to drift east and southeast away from the tanker and away from shore on the strong northwest winds. Reports from Joe Deaver and other scientists who were tracking the oil slick from the air and predicting its path with computers indicated that the heaviest concentration of the slick was passing to the southwest of the Georges Bank fishing grounds. In fact, as early as December 23, some scientists said the danger to the fishing area had passed, and others maintained there was no chance that the oil would come ashore in the United States.

Nevertheless, pollution contractors were on standby on Nantucket and along Cape Cod, ready to set out small containment booms to protect harbors, marshes, and shellfishing beds. Nantucket fishermen were particularly concerned about scallop beds in Madacut Harbor, on the west side of the island, and the main harbor, which faces north. Although both harbors faced the mainland instead of the open ocean where the slick was, the fishermen feared that currents around the shoreline would sweep the oil around the beaches and suck it into the harbors where the most abundant scallop beds are. The booms would not keep oil out of the harbors completely, but if enough of them were set up one behind the other, pollution workers could maneuver them to deflect the oil to selected sites along the shore where vacuum trucks could collect it. It was not a fail-safe system, but if it worked, the workers might be able to keep the sensitive areas of the harbors free of oil.

But that was part of the contingency plan. On December 24, with no immediate threat to the coastline, the Coast Guard, NOAA, and other groups working on the spill cut back their personnel to mimimal levels to allow as many people as possible some time home for Christmas.

Joe Deaver was not one of them. On Christmas morning, he was in the air once again charting the oil, much of which was by now more than one hundred miles away from the wrecked tanker. While his mapping flights had proved instrumental in keeping Captain Hein and other officials apprised of the progress of the slick, the information he provided scientists working with computers had resulted in some imprecise predictions of its path and left some of them doubting the accuracy of his sightings. On this flight, however, he found what he thought would put an end to the doubt: the biggest patch of oil from the *Argo Merchant* he had seen. It was a thick brownish-black blanket four hundred feet by seven hundred feet, big enough to cover four football fields. Resisting suggestions from the flight crew that he name the find after Senator Pell, Deaver called it simply "Pancake One." He plotted its position and, as the plane flew low over the huge patch, dropped a radio buoy smack into the middle of it. The buoy would emit a signal for the next day or two, making it easy for him to find the oil again and measure its precise movement for the scientists at the computers. Deaver was ecstatic. He plotted its position and then radioed the *Vigilant,* which sailed out to it to get some samples of the slick and to make a closer observation. The men aboard the Coast Guard cutter estimated that Pancake One contained a half million gallons of oil, more than six percent of the *Argo Merchant's* entire cargo.

Meanwhile, preliminary analyses of other samples of that cargo as well as some bottom samples scooped up by the *Evergreen*, the Coast Guard's research vessel, provided some more good news on Christmas Day. Dr. Jerome Milgram, a scientist from the Massachusetts Institute of Technology, who had been aboard the tanker before it broke and retrieved samples from the cargo tanks, reported to the Coast Guard that an examination of the oil indicated it could not sink unless it picked up sediments or debris that made it heavier. The *Evergreen* all but confirmed that finding when it reported that samples taken from five sites between fourteen and eighty miles from the tanker where the slick had passed showed only a trace of oil on the bottom at one site. The ocean bottom, it appeared, was being spared the impact from the *Argo Merchant*'s cargo.

But not all of the news that day was as optimistic. The winds were out of the south and southwest and, with the currents, were pushing a portion of the slick on a northerly course. On his tracking flight, Deaver also noticed some chunks of oil drifting slightly to the shore side of the tanker. He returned to the air station to combine his findings with the weather forecast, which provided information for the computer to predict the limits of the slick for the following day. It was the weather forecast that was especially disturbing. The winds were expected to shift to southeast early the next morning and blow at thirty to forty knots for up to thirty-six hours. That would carry the blobs of oil toward the coastline. The *Argo Merchant*'s slick, which had been moving so deliberately out to sea, now posed its first clear threat to the shore. Within twenty-four hours, the scientists' computer said, oil could be within two miles of Nantucket.

Captain Hein was notified immediately that afternoon and once again forces began to mobilize to combat the slick. Hein alerted pollution companies on Nantucket and Cape Cod. He asked the Army to fly one of the Skycrane helicopters back to the air station. He requested that some of the Strike Team members who had gone home for Christmas return to Cape Cod with skimming equipment and rejoin those who had remained behind. At the same time the SOR team members tried to locate some plastic drift cards they use in tracking and measuring currents that carry oil over the water. They had already used up their supply of orange playing-card size pieces of plastic on other projects, and now they needed more desperately. The cards drift on the ocean surface with the winds and currents much the way oil does, and thus provide some indication of what route a slick will follow. The scientists wanted to drop them between the advancing slick and the shore, figuring that the cards would arrive ahead of the slick and give a warning of where the oil might hit. That would give pollution workers time to set up booms and other equipment before the oil got to shore.

One of the team members telephoned Craig Hooper, the team project manager, who was in Boulder, Colorado, spending Christmas with his family, and who was close to the Environmental Research Laboratory there, where drift cards were stored. Getting the drift cards was little problem for Hooper, but getting them to Cape Cod was another matter. Shortly after he scheduled a commercial flight to Boston's Logan Airport, the airport in Denver closed because of bad weather. Hooper and the drift cards were stuck in Colorado. But the scientists and the Coast Guard officials on Cape Cod thought the threat was serious enough for Hooper to get the cards East at any cost. And after a few hours of telephone calls, Hooper finally arranged to charter a jet that would take him directly to the Cape Cod Air Station.

At 3 A.M. on the twenty-sixth, the wind over Nantucket Shoals shifted to the southeast, just as predicted. The *Argo Merchant*'s oil started moving toward shore.

The wind shift marked the edge of a large coastal storm headed toward New England, and in typical New England winter fashion, it brought a mixture of rain, freezing rain, and snow to Cape Cod that made flight conditions treacherous. Hooper and the drift cards managed to get into the air station shortly before 6 A.M., and three thousand of the cards were loaded aboard a Coast Guard HU-16E albatross, a small fixed-wing airplane that had been used for most of the tracking flights. By 9 A.M., the plane and a crew, including Deaver, were ready to go. As Deaver headed out to climb aboard in the freezing rain and snow, he wondered for a moment what he was doing there. You have to be crazy to fly in this weather, he thought. But, as important as his week of mapping flights had been to keep track of the slick, this mission was the most important of all. The drift cards could mean the difference between guesswork and a planned pollution cleanup if the oil came ashore, the difference, perhaps, between a severe environmental and economic impact on Nantucket, Martha's Vineyard, and Cape Cod and a limited one. If they didn't get off the ground that morning, the storm could shut down the operation for more than twenty-four hours. By that time, the oil would already be coming ashore.

At 9:15 A.M., the airplane shuddered down the runway and rose into the air. Within minutes, the plane was coated with a glaze of ice as it headed over land toward the southern shore of Cape Cod. Visibility was severely hampered. Deaver's knuckles were white. Out over the water, however, the temperature was slightly warmer, and at an altitude of three hundred feet the ice on the wings and windshield melted and the crew inside breathed easier.

Until that morning, the closest Deaver had seen the slick to land was about twenty-three miles. That was on the first mapping flight, on December 17, when currents had taken the first significant amounts of oil from the tanker to the west, toward Nantucket. By the next day, westerly winds were turning it back, and since then, the stream of oil flowing from the *Argo Merchant* had headed out to sea. He found the oil much sooner on this flight. The first patches showed up about nineteen miles from shore, already one-third of the way to shore from the site of the broken ship. And at 10 A.M. the winds were blowing it even closer.

The plane retreated about eight miles from the advancing edge of the slick and, at three different locations, showered most of its payload of drift cards into the water. The warning system, such as it was, was in place.

The warmer weather over the ocean made conditions good enough to continue the flight, and Deaver wanted to find Pancake One again. He had learned the night before that one Coast Guard unit had inadvertently retrieved the marker buoy he had tossed into the slick, so his radio tracking plan had been foiled. By now, however, he had become "street-wise" to the flow of currents and wind east of the tanker, and he directed his pilot back to the huge slick with little trouble. They found it about ten miles north of where it had been the day before. They dropped another batch of drift cards on the pancake, partly as an effort to track it, and partly to mark it so Deaver could identify it again on future flights. The plane then continued to chart the entire slick until shortly after noon, when they were abruptly ordered home by the air station. Weather conditions there had remained poor, they were told, and airports all along the East Coast had been shut down by the snow and ice. The air station would have to shut down soon, and the next safe landing area was in North Carolina. The pilot veered off the tracking course and headed for Cape Cod. At 1:15 P.M., they landed at the air station. The officials there immediately closed it down.

As the plane was heading home, the *Argo Merchant*'s oil slick became stalled on its path toward Nantucket. The southeast wind had unexpectedly died out. The storm had moved out to sea much further south of New England than had been forecast. By 11 A.M., the wind around the tanker had whipped back to northeast. Instead of thirty-six hours of thirty- to forty-knot southeasterly gales, the shoals had sustained only eleven hours of threatening southeast winds, from ten to twenty knots. By 7 P.M. that night, the northwest winds reached thirty to forty knots and were pushing the oil out to sea once again. The closest it had come to shore was eighteen and one half miles. The crisis had subsided.

The discovery of Pancake One meant more than precise tracking figures for the computer specialists predicting the drift of the slick. It meant the Coast Guard had a big enough patch of oil to test Senator Pell's batch of chemicals. On December 27 Deaver was asked to find Pancake One again. When he learned why, he groaned and hoped he would be unable to locate it. He objected to Pell's stubborn desire to throw the "fire dust" into the only reliable bench mark he had in tracking the spill's movement. But he found the large slick with ease and escorted a helicopter out to try the experiment. Crewmen from the helicopter dropped several boxes of the wicking agent and some aviation fuel into the slick. The chemicals were ignited by a delayed fuse. Within minutes, thick black smoke rose from the slick as the mixture caught fire. Flames leaped fifteen feet off the surface of the gunk. The fire lasted about forty-five minutes, long enough for Pell to be delighted and short enough for the Coast Guard to call it a failure. Just how much, if any, oil actually burned was never conclusively determined. But when Deaver flew over the slick after the flames were extinguished, he knew he still had his pancake. Its size and shape had not changed. He plopped another marker buoy into the goo to help him locate it the next day, and went home.

For the next three days, bad weather and a bad airplane grounded the tracking crew. But the northwest wind kept blowing, moving the outer edge of the oil more than 130 miles from the wreck site and about 160 miles from Nantucket. Scientists following the slick believed that the threat to New England shores by now had really passed.

The "threat" to Pancake One was not over yet, however, and on December 31 Deaver took off once again to escort a burning party to the oil. This time, the Coast Guard was trying the experiment from the buoytender *Spar*, which had been called back to the incident. The patch was smaller than it had been on the twenty-seventh, and with the turbulence caused by the maneuvering of the *Spar*, Deaver watched from aloft as it broke up into what he called pancakes sub two, sub three, sub four, and sub five. The men on the *Spar* chose a thirty-by-sixty-foot blob, and tossed several eleven-pound bags of the wicking agent into its middle. Some of them ripped open on impact; others were torn open with birdshot from a shotgun fired on the vessel. The powder in the bags had the consistency of cigarette ash, and the wind immediately blew it into the water. Before they quit, the men tossed enough wicking agent onto the slick to cover it with a pile one foot high. But ninety-five percent blew away. Then they coated the slick with sheets saturated in jet fuel. Just one of the sheets caught fire when it was ignited with a flare, and it burned for four minutes. The experiment was called off. Pancake One was destroyed, but it had not burned.

After finding themselves grounded for three straight days, the SOR team decided to end the hit-or-miss tracking of the slick by dropping a large floating marker buoy which sent out radio signals not to shore or to an airplane but to a satellite. They planned to continue their flights as long as Captain Hein needed them, but the satellite would ultimately take over responsibility for tracing the path of the *Argo Merchant's* cargo. And there was little doubt at year's end that it would be tracked out to sea.

18

The Aftermath

While most people ashore were preoccupied with the oil floating on the water, Barry Chambers and the Strike Team were still concerned about what was left of the *Argo Merchant*, as well as any cargo that might remain trapped in her tanks beneath the surface. On Thursday, December 23, the day after the long, silent helicopter ride back to the air station, Chambers decided to go aboard the bow section that was riding high out of the water. He wanted to inspect what tanks he could for oil, and he especially wanted to open every hatch he could reach so the holds would fill with water and the bow would sink right where it was. He already knew the number one port tank was empty, and thus full of air. That gave the hulk enough buoyancy to float for a long time, he figured, and no one wanted it to drift away in the currents and the surf. Once it was sunk and stable in the sand, he and some of his divers on the Strike Team could dive on it to check all of the tanks. Chambers also thought that sinking it immediately would give the *Vigilant*, which was under orders to watch the tanker's remains, a chance to get home for Christmas. He briefed his team on the idea.

"Okay, I need two volunteers. Mac?"

"I'm goin'," McKnight said.

"Klink?"

"Ya, I'll go," Klinefelter replied.

"Okay," Chambers said. "Now, the bow is sticking up in the air a little ways and we're going to suit up in the Unisuits, get some repelling lines, climb out on the bow, and sink the son of a bitch."

A Unisuit is a one-piece wet suit that covers the body from neck to toe and insulates it against cold water. The Strike Team routinely uses them on diving jobs in cold weather or when they are likely to get wet while they work. This wasn't a diving job, but Chambers expected to get wet. In fact, he expected to end up in the water. McKnight and Klinefelter didn't know that. They thought the job sounded routine.

The men pulled on the wetsuits, gathered some rope and some tools to open the hatches. Shortly after noon, they climbed aboard the helicopter and headed out to the tanker once again.

On previous trips to the tanker's deck, the Strike Team had been lowered out of a helicopter in a metal rescue basket. This time, however, they were going down in a horse collar, a leather strap that loops around a person's chest under each arm and suspends him from the end of the helicopter's cable. Klinefelter had never used one before. In fact, the previous Sunday was his first trip in a helicopter, and he was not well versed in stepping out of them in midair. He did know that to get out of the horse collar, all he had to do was stick his arms straight up in the air and he would drop through the loop. He was a good athlete, strong and agile, and he didn't think that sounded hard at all.

Klinefelter had second thoughts, however, when the helicopter arrived over the tanker. The bow was not just sticking up a little ways. It was sticking up at a sixty to seventy degree angle.

"That's almost standing straight up," he mumbled as the three of them looked down on the bow, which also was listing sharply to starboard. He wondered how he would land on it. Except for the tank covers, the anchor windlass, and a few other fixtures that protruded off the deck, the bow was as sheer as a cliff. Most of it was coated with oil. Klinefelter was not alone in his surprise. McKnight glared at the bow, then looked over at Chambers.

"Are you shittin' me?" he said.

Chambers grinned.

To make sure he had enough power to hover safely while the Strike Team was lowered to the ship, the pilot first dropped McKnight and one of the crewmen off on the *Vigilant* to lighten the load. Back over the slanting bow, Klinefelter sat at the edge of the cargo door, his feet dangling in the air, and strapped the horse collar around him. The other helicopter crewman hooked him to the cable from the winch. The noise of the engine and the blades made it impossible to hear anything in the cabin, so the crewman gestured to Klinefelter to see if he was ready. Klinefelter nodded. The crewman pushed him out the door. For a split second, Klinefelter plunged toward the water and the ship. Then the cable pulled taut. The belt tightened under his arms and jerked him to a stop. He hung there in the air for a moment, suspended just below the whirring rotor blades. His pulse raced.

"Christ," he said aloud to himself. "I didn't know they just threw you out of the plane."

He felt himself descending toward the ship about fifty feet below him. When he was halfway there, however, he suddenly felt himself going back up again. They pulled him back into the cabin, flew around for a few more minutes, and then threw him him out of the helicopter again. This time he dropped quickly toward the ship. He hit the tip of the bow, kicked off with his feet, and swung away. He swung back, slammed against the face of the bow on his back side and slid down into the anchor winch. He grabbed it. The crewman gave him some slack and he slipped out of the harness. As he watched it go back into the air toward the helicopter, Klinefelter took a gulp of air and let it out slowly.

Chambers was next. Part way down, he got caught in the turbulence of wash from the rotor blades. He started to spin. Chambers put his feet out. He kicked frantically in the air. He jerked his body back and forth. Nothing worked. He could not stop the spin. He kicked his feet out searching for the deck. As soon as he felt the touch, he threw his arms into the air and dropped out of the horse collar. He fell against the deck and slid hard into the anchor windlass. The angle of the deck and the oil made it difficult to stand, but Chambers didn't even try. He was dizzy. The bow and the water were spinning around him.

The helicopter went back to the *Vigilant* to pick up McKnight. He was lowered to the port side edge of the bow, and slid down against the windlass, where Klinefelter braced his impact.

"Nice day out here, huh?" Chamber remarked, having regained his balance by the time McKnight arrived.

It was sunny and not too windy, but McKnight just frowned. He looked over the starboard rail, where he would have gone if he hadn't slid against the anchor windlass. Down in the water, he saw the smokestack and the upper deck of the afterhouse of the tanker, still poking out of the water, awash with white foam. McKnight also felt the bow swaying back and forth in the three foot swells. With the sharp starboard list, McKnight thought sure the bow was going to roll right over and throw the three of them right on top of the smokestack.

"Okay, we're gonna take these repelling lines and tie them to the rails and hang on to 'em while we open up these hatches," Chambers said. "And I want you to pick a spot to go off her if she decides to roll over."

The *Vigilant* and one of the tugboats were nearby. McKnight figured they could swim toward the *Vigilant* if they had to go into the water. He just didn't want to get wedged in between the bow and the stern if she did start to flip.

The men moved carefully about the fo'c'sle, the triangular area at the tip of the bow raised one deck above the tank deck. They pulled themselves up the deck on lines to reach the uppermost hatches, and worked their way around the rail, opening anything they could find that covered a passageway to a hold that could fill with water. As they worked, the constant surging of the bow gave them a topsy-turvy feeling and had the effect of throwing them toward the starboard side. Several times, the men lost their footing completely. Their grip on the lines pulling against their own weight was all that kept them from slipping into the sea.

Their biggest job, however, was down on the main deck, at

the battened down tank cover on the empty number one port tank. The men tied a line off to the windlass and repelled down the fo'c'sle to the ladder on the port side. They looked down on a small section of the main deck that emerged from the water. It was covered with three inches of oil. And only the number one port tank was within reach. The others were awash. The men would not be able to check them to see if any cargo remained inside.

They eased themselves down the ladder and crawled through the goo. It was not possible to stand up. Chambers boosted himself onto a deck pipe and wrapped his legs around it to hold on. He grabbed one end of a rope that was tied around Klinefelter's waist and hung on as Klinefelter crawled up the deck to the tank cover. McKnight got between Chambers and Klinefelter and watched the waves coming in against the ship.

"Stand by," he hollered to Chambers. "You're going to get hit."

Chambers, who was closer to the water than the others, hunkered down against the pipe. The wave crashed over him. He was completely buried in the wash. The wave receded and Chambers reappeared, his face blotched with oil.

"How's the water!" McKnight hollered.

They all laughed.

Chambers was buried once more before Klinefelter raised the top on the number one port tank. The men also opened hatches to dry storage areas just forward of that tank. McKnight noticed that the portside door, leading to a large hold under the fo'c'sle, was dogged shut.

"We oughta get this door up here," he said, moving toward it on his belly, slithering through the oil. He got as far as the base of the ladder he and the others had come down a few minutes earlier. He could get no further.

With McKnight lying there, however, Chambers got an idea. They couldn't get any footing on the oil covered deck, but maybe they could get some on each other.

"Awright, Mac, dig your feet in good around that ladder. I'll climb over you and lay down above you, then Klink, you climb over both of us, reach up and open that door."

Chambers pulled himself through the oil, climbed over McKnight and stretched out beside him in the gunk. Klinefelter followed, crawling up the human chain.

McKnight felt Klinefelter's feet digging into his back. He felt the squish of oil under his already-oiled wet suit. He wondered what the hell he was doing lying prone on the slanting remains of the *Argo Merchant* while someone else stood on his back. He glanced over at Chambers lying beside him.

"You remember that time I told you I'd go anywhere with you?" he said.

"Yeah."

"Well, we're here."

Klinefelter could just reach the bolted latches that kept the door shut. He flipped over a latch at the top of the door, then reached down and pulled the one on the bottom. The door swung open.

"Okay," Chambers said. "Let's get the hell out of here."

"Good idea," McKnight said slowly with a forced grin.

Chambers radioed for the helicopter, which had landed on the *Vigilant*. Then the three of them pulled themselves up the port ladder to the fo'c'sle where they waited for a ride home.

McKnight grabbed the horse collar when it came down the first time. Chambers went up next, leaving Klinefelter alone on the swaying deck. With no one to talk to and nothing to do but wait, Klinefelter got jittery. Each movement of the bow seemed more pronounced. He was sure it was going to roll.

"Hurry up, dammit, hurry up and get here," he said aloud as the horse collar came toward him in what seemed to be slow motion. "If this son of a bitch turns over, I've had it."

The strap finally reached him. He grabbed it, draped it under his arms, and gave the signal to the crewman in the helicopter. The cable tightened. He felt the pressure of the strap under his arms. His feet lifted off the deck.

"Aw shit, man," he said, with both jubilation and relief. "That's done."

Jim Klinefelter was the last person to set foot on the *Argo Merchant*. The bow didn't sink immediately, however, because a storm that had been forecast never came. Instead, the following three days were the calmest the Strike Team had seen since they had arrived. Thus, in the end, the last visit aboard the tanker accomplished little. The men were unable to determine if any oil remained in the forward tanks. And the *Vigilant* was forced to remain on duty through Christmas to keep a watch on the bow.

At first, the bow didn't go anywhere. The heel of the forward section was either still attached to the submerged bridge section or it was resting on the sand. But over the next five days, the bow worked itself free. It did not level out in the water. The angle of the deck got steeper and steeper. Without the unusually heavy winds and seas of the previous week, the currents on the shoals worked alone on the bow, with patience and incredible strength. On the morning of December 28, the *Vigilant* found the bow capsized and floating in the water. But it had not rolled over to starboard, as the Strike Team had thought it would. The currents, working underneath the hull, pushed and pushed until the bow stood up straight in the air and finally fell over backwards. The power of the water on that shoal just flipped it over.

Coast Guard officials didn't want that to happen. The air pocket was trapped within the bow, and all the hatches that Chambers, McKnight and Klinefelter had opened were underwater. The air couldn't escape. The *Vigilant,* after two weeks of constant surveillance and support operations around the tanker, was relieved by the *Bittersweet* on December 28. Three days later, the *Bittersweet* used her twenty millimeter deck gun and shot holes in the bottom of the bow. Even with the air pockets broached, it was six days later, January 5, before the Argo *Merchant*'s bow finally disappeared below the surface.

On January 8, Chambers and Don Miller, another member of the Strike Team, returned to the scene of the grounding in a tugboat to dive down on the broken ship and check her holds for oil. While northwest winds were carrying what oil had escaped away from shore, any oil that might remain in the tanks would act as a time bomb that could go off anytime, perhaps in the summer when prevailing winds would carry the oil toward crowded beaches. Just one tank on the *Argo Merchant* contained as much as 250,000 gallons of oil. No one wanted to be surprised by a slick like that coming ashore in July.

When Chambers and Miller got into the water, however, they couldn't dive on the bow section. The current was too strong. They couldn't get down on the stern section, by the smokestack and engine room, either, for the same reason. The five- to six-knot current threatened to carry them away. They managed to get down on the midsection, however, by pulling themselves down one of the kingposts that still poked out of the water. When they reached the bottom, both Chambers and Miller were stunned. Chambers had been impressed by the strength of the water on that shoal again and again: on the Monday before the ship broke, when the current foiled every effort to get the anchors out; when he flew over the ship after it had broken the second time; and when he learned that the current had picked up the bow section and flipped it over. Still, he was not prepared for the sight that was in front of him now. They found the main deck of the ship. Under the deck was sand. No tanks. No hull. The ocean had ripped the deck right off the the ship, like peeling the top off a sardine can, Miller thought. Inch thick pieces of steel were torn like shredded paper and mangled. Cargo pipes were bent and twisted like pretzels. And pieces of metal were strewn all over the bottom, as if a bomb had exploded inside the vessel. To ask if any oil remained seemed laughable. The *Argo Merchant,* Chambers thought, looked like a child's plastic boat that had been stepped on and crushed.

By early January, Captain Hein was anxious to phase down the extensive and expensive pollution tracking and cleanup measures that had been in effect since shortly after the grounding. He released the pollution contractors who had been standing by at Chatham, Massachusetts, since Christmas Day, when the oil was forecast to move toward the coast. And he announced plans to let the private pollution crews on Nantucket go home as well. Island officials, who finally got a liaison person to work with during the latter part of December, agreed with

Hein that the slick posed only a remote threat to the island and that oil forecasts would provide plenty of time to remobilize pollution workers before any oil came ashore.

But Massachusetts officials, particularly Lt. Gov. Thomas P. O'Neill 3rd, and Evelyn Murphy, the Secretary of Environmental Affairs, objected. They had been especially critical of the Coast Guard's handling of the whole incident from the beginning, and they insisted that Hein's plan was premature. He agreed to keep the crews on standby for a while longer.

Meanwhile, with weather conditions frustrating efforts to dive on the bow and stern sections, Chambers and other officials devised a plan to blow up the bow, though they knew it could contain as much as 1.3 million gallons of oil. Since it had broken from the midsection between the number three and number four tanks, it was almost certain those tanks had been ruptured. That still left number one center and all three number two tanks unchecked. If any oil were in them, the officials thought it would be better to release it now, while the winds were blowing from the northwest, rather than later when it might come ashore. In announcing the idea, the Coast Guard said it would check the bow for oil before blowing it up. The plan met stiff opposition from fishermen and environmental groups when it was aired at a public hearing in Falmouth, Massachusetts, on January 19. Two days later, O'Neill and Murphy expressed their opposition to the move to the National Response Team in Washington. That team, which oversees the national oil spill contingency plan, urged the local Coast Guard officials to inspect the bow before making a decision on the plan. At the Bay State officials' request, the national team also suggested that the Coast Guard continue to maintain pollution contractors on Nantucket until the Strike Team determined just how much oil, if any, was trapped in the tanker's remains.

During the first month of the intense activity around the *Argo Merchant* incident, so many other tanker accidents occurred in or near U. S. waters that they seemed to lay siege to the coastline. And most of the ships were Liberian.

On December 17, 1976, the *Sansinena,* an 810-foot Liberian

tanker, owned by the same company that owned the *Torrey Canyon*, blew up in Los Angeles Harbor. Nine persons were killed, fifty-eight were injured, and twenty thousand gallons of oil spilled into the harbor. The ship, valued at $21.6 million, was a total loss.

On December 24, the Liberian-registered tanker *Oswego Peace* spilled five thousand gallons of oil into the Thames River, near Groton, Connecticut, after striking a submerged object in the Delaware River enroute to Groton, cracking her bow. The spill polluted a 3,600-foot stretch of beach along the river.

On December 27, the Liberian-registered *Olympic Games* ran aground in the Delaware River, near Philadelphia, spilling 133,500 gallons of oil, which coated the beaches of three states, fouled fragile wetlands, and killed hundreds of birds.

On December 28, the Liberian-registered *Daphne* ran aground in Guayanilla Bay, Puerto Rico, without spilling oil.

On December 30, the *Grand Zenith*, a Panamanian tanker with a crew of thirty-eight and a cargo of 8.2 million gallons of oil, vanished in the North Atlantic, northeast of Cape Cod. A twelve-day search by the Coast Guard turned up a few bits of debris, no oil and no crewmen.

On January 4, 1977, the Liberian-registered *Universe Leader* ran aground in the Delaware River, near Salem, New Jersey. It did not spill oil.

On January 5, the United States-registered tanker *Austin* spilled 2,100 gallons of oil into San Francisco Bay while unloading at Martinez, California.

On January 7, the Liberian tanker *Barcola* ran aground twenty-five miles off the Texas coast, on its way to Port Arthur. No oil spilled. The same day, an explosion aboard the Liberian tanker *Mary Ann*, three hundred miles off the Virginia coast, resulted in injuries to two crewmen. The ship was cleaning its tanks when the explosion occurred.

On Janaury 9, the barge *New York* spilled 42,000 gallons of diesel fuel when one of its tanks ruptured in the Sparkman Channel, Tampa Bay, Florida.

On January 10, the United States tanker *Chester A. Poling* split apart in heavy seas off Gloucester, Massachusetts. One crewman died and six others were saved in a dramatic rescue when they were plucked from pieces of the ship in rescue baskets lowered from Coast Guard helicopters. The ship was not carrying oil.

On January 17, the Greek tanker *Irenes Challenger* broke in half 220 miles southeast of Midway Island in the Pacific Ocean. Three crewmen were lost and 3.2 million gallons of oil spilled into the sea.

On January 19, the Liberian tanker *Golden Jason*, carrying 9.2 million gallons of oil, was towed into Newport News, Virginia, after suffering a breakdown at sea. The ship had been sold for scrap but then got a contract for one last cargo of oil before heading off to the cutting torch.

Some members of the Strike Team responded to some of those incidents, but Chambers remained on Cape Cod. At times, he felt stranded there. He figured the bow and stern sections had fared little better than the mangled midsection he had already seen. But he still had to check them out to make sure, and the weather and currents along Nantucket Shoals were not through with him and his team yet.

After waiting out several days of bad weather, public hearings, and planning sessions, Chambers, McKnight, Klinefelter, and Miller hired a sixty-five-foot tugboat which took them out to the wreck site on January 20. But icy weather and heavy seas made it impossible to work from it. They returned to port three days later. The team never got into the water.

The regional response team in Boston arranged to use the Navy's *USS Recovery* as a diving platform ship. She arrived in Newport, Rhode Island, January 27 and left for the tanker later the same day. The next morning, the Strike Team launched their twenty-one-foot rubber boat called a Zodiac. Using a depth recorder, they picked up a signal they thought was the bow near where it had sunk three weeks earlier. But the current dragged them away from the spot before they could drop a marker, and before they could find the spot again, the seas picked up. They halted their operation and, in the face of a severe winter gale, the *Recovery* headed back to Newport for shelter.

The storm lasted three days, and it was February 1 before the Strike Team and Navy divers returned to the wreck site on the *Recovery*. In the wake of the storm, they noted that a marker buoy Chambers had attached to the kingpost bobbing out of the water had disappeared. They also found that a large navigation buoy that was anchored with a ten thousand-pound clump of concrete had been carried three miles south of the wreck. They spent the next three days looking for the bow. But it was no longer where they thought it had gone down. Chambers asked if a minesweeper, with intricate bottom-sounding machinery, could be brought in to assist in the search.

Meanwhile, on February 4, they finally dove down on the stern of the tanker. Like the midsection, it was completely ripped apart. The entire deck house, including the smokestack, was destroyed. Pieces of the engine room, including pump motors, condensers, steam coils, and valves were scattered like shells about the ocean bottom. No oil tanks were visible anywhere. In a departure from the usual stilted and dry officialese of a Coast Guard message, Chambers informed officials ashore that "the level of destruction was incredible to those who saw it and a physical testimony to the forces on this shoal."

The next day, Chambers and his men searched an area about one and a half miles from the wreck site where an airplane pilot flying over the area thought he spotted a sheen of oil. They found what they believed was the missing bow section. If they were right, it meant that the gale during the final three days of January had hit with such force that the submerged bow section had tumbled and dragged more than 2,600 yards along the sandy bottom.

Chambers immediately dropped a marker with an anchor on it into the water, and his men prepared to make a dive. It was snowing hard and it was cold. The wind and freezing spray fouled their diving gear. By the time they got into the water, the current had swept the marker away from the site. When the divers reached bottom, they found nothing but hard white sand. They surfaced into the teeth of yet another advancing storm. And for the third time in two weeks, Chambers and his men returned to shore empty-handed to wait out the weather.

The *Recovery*, along with the Minesweeper *Dash*, returned to the scene once again on the morning of February 9. Chambers and his men were tired and frustrated. McKnight had managed a quick trip home for Christmas, but none of the others had been home since December 15. Chambers felt like a permanent resident of Cape Cod. Diving in the freezing cold and the swirling currents was not sport. The men risked their lives each time they went overboard. Chambers considered the operation more dangerous than anything his men had done aboard the tanker and he was certain it was also more futile. Thus, the men were encouraged that morning when the *Dash* picked up metallic soundings almost immediately in the area one and one-half miles from the wreck site where Chambers thought he had found the bow a few days earlier. The minesweeper dropped several anchored markers into the water and the Strike Team and Navy divers headed for them in two Zodiacs. By the time they got in the water, however, the currents had once again moved the markers. The divers hit bottom and found nothing. Despite clear diving conditions, they were unable to venture from the lines of the marker anchors. The currents were too strong for them to search a wide area. Miller tried digging his hands into the bottom and pulling himself along. It was impossible.

The men made four dives that day, and they did find a four-foot by ten-foot piece of the tanker's hull. But it wasn't the bow.

At 9 A.M. the next morning, the *Dash* picked up the bow once again. This time her crew littered the spot with markers. Once again, divers working out of the Zodiacs jumped into the water. Miller and McKnight were the first ones down. This time the markers were right on target. The bow section lay on the bottom upside down. Her hull was marred by gashes from the gunfire that had led finally to its sinking.

"That's the prettiest thing I've seen in a long time," Miller said to himself as he headed topside to inform Chambers. He poked his head out of the water, ripped his mask from his face, and looked over at Chambers in the Zodiac.

"You buy the beer," he said. "We've got her now."

Chambers smiled. "It's about time."

Chambers jumped in the water with Miller and two Navy divers and went down to see for himself. The bow was not as severely damaged as the other sections of the tanker. The main deck that covered the first three rows of tanks was torn away, but tank bulkheads were intact. Chambers swam into one of the tanks and found gobs of oil clinging to the sides. While some of the clingage would seep out and rise to the surface in droplets and produce a sheen, he found no significant pockets of oil that would gush from the hulk if heavy currents or gales flipped it over again. For practical purposes, the *Argo Merchant* was empty. The Strike Team, smiling broadly, headed back to the beach for the final time.

With Chambers' report, Hein phased out the special operations set up for the incident. Pollution contractors standing by on Nantucket were sent home. Coast Guard beach patrols on the island were ended. Special weather forecasts for Fishing Rip were no longer needed. And scheduled daily flights by Coast Guard aircraft were canceled. Instead, the site would be checked during the air station's normal twice-weekly pollution surveillance flights along the coast.

On February 11, fifty-nine days after the *Argo Merchant* ran aground, the Atlantic Strike Team was also released from the project. Chambers and his men went home.

Epilogue

On February 12, 1977, and again on February 27, scientists aboard the *Endeavor*, the scientific research vessel of the University of Rhode Island, reported finding significant concentrations of oil in bottom sediments near the *Argo Merchant*. The information contradicted the earlier findings based on divers' observations of handfuls of sand. But the scientists were not finding large globs of goo on the bottom. These were tiny droplets of oil sticking to the appendages and the digestive tracts of animal plankton, the microscopic organisms that make up the base of the marine food chain. They were found in the digestive tracts of two crabs, one dead and one nearly dead. And they were found in the sandy sediments of the shoals.

Other cruises, by the National Marine Fisheries Service, established that fish eggs—especially cod and pollock, which were breeding during the early winter—were killed in massive quantities in the waters immediately around the wreck site. Research later established that mussels and other shellfish suffered impaired respiration.

But although the slick itself extended at least two hundred miles from the vessel by mid-February, the significant traces of oil were found within a twenty-square-mile area of the tanker. Any oil that did find its way into the water column or to the bottom was apparently widely dispersed and diluted by the currents, winds, and waves beyond that small patch of sea.

About 150 birds were picked up along the shores of Nantucket and Cape Cod. Ninety-one were taken to the Felix Neck Wildlife Sanctuary on Martha's Vineyard; the remaining fifty-nine were dead. Of those taken to the sanctuary for treatment, forty-four were either dead on arrival or put to sleep immediately; twenty-two died during the cleansing and rehabilitation treatment. About twenty-five birds were released after treatment.

The number does not represent all those that washed ashore nor does it include offshore birds which may have been affected by the spill. But the impact on birds is believed to have been minimal.

Other than a light sheen caused by oil seeping from the sections of the wreck, the last day any of the pilots and scientists actually spotted the tanker's slick was January 28, 1977. After that, the pancakes were too small and dispersed over too wide an area of ocean to be detected. Scientists determined that the action of the water compressed and compacted the pancakes, making them smaller and heavier and eventually transforming them to tarballs, which probably have drifted about the sea with the currents. Where those tarballs went is anybody's guess, but at the end of 1977 the satellite buoy tossed near the remains of Pancake One was still adrift, moving aimlessly about a three thousand-square mile patch of water just west of the Azores, more than three thousand miles away from the Nantucket Shoals.

In March, 1977, the preliminary findings of the scientists who investigated the effects of the spill or followed its path were published in a scientific report compiled and edited by Dr. James Mattson and Dr. Peter Grose, of the Spilled Oil Research team. Despite the initial bickering and lack of coordination among the researchers, the report was widely regarded as a thorough early analysis of the impact of the spill. Most of the scientists concurred that while definitive statements were impossible to make, the *Argo Merchant* oil spill was not the major environmental catastrophe that many had predicted it would be. While a small area of the sea was clearly injured by the oil in ways that may never be known, considering the magnitude of a seven-and-one-half-million-gallon oil spill, the impact was minimal. Another year of research into the spill turned up nothing to alter that conclusion.

Meanwhile, the scientists have taken steps to cure the organizational ills that were so obvious during the first several days on Cape Cod. At the beginning of 1978, researchers for the Environmental Protection Agency and the National Oceanic Atmospheric Administration were working with other scientists to set up an environmental strike team that would respond to significant spills and coordinate the scientific effort to track them and determine their impact.

On January 11, 1977, Sen. Warren Magnuson, Democrat of Washington state, called a hearing of the Senate Commerce Committee to sound off against the crippled armada of tankers that were entering U. S. ports, to develop legislation aimed at alleviating the problem, and to vilify the Coast Guard for not using what authority it had for keeping substandard vessels out of U. S. waters.

During those hearings, Coast Guard officials, including Admiral Owen W. Siler, Commandant, explained that the Coast Guard checked foreign vessels on a spot basis, usually limited to inspections of cargo loading and unloading procedures unless conditions aboard a ship were flagrantly bad. They said the inspection procedure was in line with an international agreement known as SOLAS, the Safety of Life at Sea convention. Under it, any vessel which holds an up-to-date SOLAS certificate is assumed to be in good condition and not subject to domestic regulation of nations who have signed the treaty.

But with foreign flag tankers leaking, grounding, and exploding up and down the coast, and with the Congress clamoring for action, the Coast Guard on January 21, 1977, launched an ambitious inspection program, intending to board and inspect every foreign flag tanker that entered U. S. waters. The move was prompted specifically by the explosion of the *Sansinena* in Los Angeles harbor. In its investigation of that incident, the Coast Guard determined that the vessel's cargo venting system, which was to carry volatile tank vapors harmlessly into the air, was full of holes and leaking vapors onto the deck, exposing the innards of the tanks to any source of fire—from sparks caused during ship maintenance to the lighting of a cigarette—on the main deck. The Coast Guard also learned that the shipowners' classification societies, such as Bureau Veritas, which surveyed the *Argo Merchant*, were not paying much attention to the cargo vents. The Coast Guard decided it had better do the inspections itself.

In the first year of the program, they inspected 2,650 foreign tankers. More than fifty percent violated safety standards. In all, the inspectors found more than eight thousand violations. One of the vessels they inspected was the *Agia Erithiani*, a 631-foot Liberian tanker laden with 31,500 tons of number six fuel oil, about four thousand more tons than the *Argo Merchant*. She had reported suffering damage in rough weather en route from Teesport, England to Fall River, Massachusetts. Upon boarding her, Coast Guard inspectors found twenty feet of oil in the forward pump room, which is supposed to be free of oil so the pumps can operate during loading and unloading. They found four feet of oil in a dry storage hold in the bow of the ship, forward of the pump room. They found three and a-half feet of water in a chain locker, another dry compartment in the bow. They found three and one-half feet of molasses and water, apparently the remains of an earlier cargo, in the after pump room, between the last row of cargo tanks and the engine room. And they found life boats that were damaged and unusable by the ship's crew in case of emergency.

A ship encountering rough weather might be expected to have water in the chain locker, which has vents open to the deck that could collect water that sloshed over the bow. But hundreds of gallons of cargo floating around outside the cargo tanks is not routine storm damage. The *Agia Erithiani* was not permitted to sail on to Fall River. The Coast Guard ordered her to remain at anchorage off Jamestown, Rhode Island, until the pump rooms and dry compartments were cleaned of oil and temporary repairs were made so she and her crew could proceed safely to port. The repairs took two weeks.

Six months into the program, the Coast Guard added navigation equipment to the inspections, which had previously centered on cargo-handling machinery and procedures. And throughout the program, they recorded each boarding and each safety deficiency and stored the information in a computer. Many of the smaller, aging tankers worked almost exclusively in United States trade, and return to American ports frequently. Whenever they do, the local Coast Guard port officials can check the computer for the vessel's boarding and inspection history and know in advance whether they are dealing with a troublesome ship. In most cases, vessels found in violation of safety regulations on their first inspection have had fewer violations by the third and fourth time they enter a U. S. port. Thus, the Coast Guard believes the program quickly proved effective in improving shipboard conditions on these vessels. How effective it will be in reducing tanker accidents and oil spills is a question first-year statistics could not answer.

In March, 1977, both Senator Magnuson's Commerce Committee and President Jimmy Carter issued a series of proposals designed to reduce the risk of oil spills. They included major improvements in tanker design and construction standards, tank cleaning processes, and navigation equipment and practices. Many of the regulations were to apply to old as well as new tankers over twenty thousand deadweight tons to protect American ports from the size of vessel which trades in them. Carter also asked for studies and research into oil spill cleanup equipment and procedures aimed at improving the nation's ability to respond to large oil spills.

The proposals also included the establishment of a $200 million "superfund" which would provide money for cleanup for oil spills, for damages suffered by victims of such spills, and for certain research into their effects. The federal government would pay these expenses, then go after the people who spilled the oil to get the money back. In the *Argo Merchant* incident, the Coast Guard had access to a much smaller federal "pollution" fund, from which Captain Hein spent about $1.8 million in the two-month effort. In October, 1977, the government recovered $1.1 million after negotiations with a group known as TOVALOP, tanker owners voluntary agreement for liability for oil pollution, of which Thebes Shipping, Inc., was a member.

The legislation growing out of these proposals was, for the most part, held up in Congress at the end of 1977 while leaders awaited the reaction of the international community through the Inter-governmental Maritime Consultative Organization, a United Nations group more commonly known as IMCO. The rash of oil spills plus pressure from the United States caused IMCO to accelerate its schedule for considering tanker construction and navigation and watchstanding agreements. If strong standards were not adopted by international agreement, Senator Magnuson, Rep. Gerry Studds, and other legislative leaders were ready to make sure that the United States, with strong legislation of its own, took action unilaterally.

A crisis usually inspires strong action, and the proposals and programs put forward to improve the nation's ability to handle oil spills are the positive results of a seven-and-one-half-million-gallon oil spill.

"I don't think anything but good came out of the *Argo Merchant*," Barry Chambers remarked several months after the incident. "The *Torrey Canyon* did not occur in this country, but it had a significant impact on the environment in this country. The *Argo Merchant* did occur in this country and it continued the impact of the *Torrey Canyon*. But fortunately, it had no significant environmental impact, so that we didn't have to have a disaster to make us face up to some of the problems that we do have. The end benefactor in essence is the environment."

In the months after the spill, some environmentalists and fishermen caught themselves wishing the *Argo Merchant*'s cargo *had* come ashore, figuring that a disaster would have driven the point home much deeper. It is important that the lessons of the *Argo Merchant* remain in the forefront of the public's mind even though the oil from her holds has long since disappeared. For the *Argo Merchant* and the accidents that followed were not an aberration. They provided a vivid and dramatic display of a chronic national and international problem.

At any one time, oil industry representatives say, forty-two billion gallons of oil are floating in tankers on the oceans of the earth. The United States in 1977 was importing daily the equivalent of more than forty *Argo Merchants* full of oil, much of it in marginally operated old tankers like the *Argo Merchant* itself. The spills during the month following that incident were among more than ten thousand spills a year reported to the Coast Guard. While most of those are minor spills, such as the ten- to sixty-five-gallon leaks the *Argo Merchant* suffered during the last few years of her life, some are clearly much more severe.

"Much is at stake here," said Sen. Ernest F. Hollings, Democrat of South Carolina during the Senate Commerce Committee hearings in January, 1977. "While the oil that tankers bring in is absolutely vital to our well-being, we can no longer tolerate such major environmental damage, and such serious economic threats to our coastal areas and ocean fisheries."

"The series of catastrophes," added Sen. Edward W. Brooke, Republican of Massachusetts, "has humbled us all by demonstrating the vast and frightening power of the oceans and our relative ignorance and helplessness in the face of these forces."

Indeed, the *Argo Merchant* revealed our helplessness. The battle mounted against that tanker to prevent the spill was unprecedented. More than one thousand men, including the leading oil pollution and salvage experts in the country, spent hundreds of thousands of dollars and worked with millions of dollars worth of the most sophisticated oil spill prevention and cleanup equipment available anywhere. A week after the ship went aground, that effort ended in failure. Except for the goo that was smeared on Strike Team workers' hardhats, that stuck in their hair, clogged their fingernails, and stained their coveralls, not one drop of oil was saved before the *Argo Merchant* broke apart and spilled her cargo into the sea.

But the same northwest winds and vicious currents that

foiled that effort also carried most of that oil out to sea, sparing the coastline and the lucrative Georges Bank fishing grounds from significant environmental damage.

"We were damn lucky," Barry Chambers said. "That's all it was."

Author's Note

This book is based on information I gathered from more than fifty interviews, from more than four thousand pages of testimony, and from a variety of published sources. I was unable to interview the captain and crew of the *Argo Merchant,* but they told their stories in substantial detail in depositions in U.S. District Court in New York City and before a special inquiry held by the Liberian government. Chapters I and XVI are reconstructed from that testimony. Conversations depicted in those chapters are not actual word-for-word renditions, but are reconstructed according to the men's recollections of their conversations as related in their testimony.

In addition to my own reporting, the information in Chapter XV on the history of the *Argo Merchant* is compiled from Congressional testimony, news articles in the *Providence Journal, The New York Times,* the *Wall Street Journal* and *Audubon* magazine, and a report prepared by Rep. Gerry Studds of Massachusetts. The details of the *Argo Merchant*'s accident record are based on information compiled by Arthur McKenzie, of the Tanker Advisory Center in New York, and on Coast Guard records.

The rest of the book is based mainly on information gleaned from interviews with the participants in the effort to save the tanker and to prevent and then track the spill. Conversations are reconstructed according to what the men recalled of their conversations during the effort. Most of the major characters have reviewed portions of the manuscript to verify technical information and confirm the accuracy of the conversations as to context and content.

Those who followed the incident closely may remember that maritime experts had speculated that the captain and the chief officer were getting a reciprocal reading on the radio direction finder during the final hours of the tanker's last voyage. The instruments are such that if not used carefully, they can give a reading 180 degrees from the actual source of the radio signal. For this to have occurred in the *Argo Merchant*'s case, the Nantucket Lightship would have had to have been directly off the stern of the ship during the last ninety minutes of her the voyage. A reconstruction of her voyage, however, based on her course and her officers' testimony, indicates the Lightship was to the tanker's starboard side during that time and never directly astern. Thus her captain and first officer could not have had a reciprocal reading.

Other observers of the *Argo Merchant* incident have speculated that her captain was purposely taking a shortcut through dangerous waters to the shoal side of the lightship to make up for lost time and to arrive in Salem by the 5 P.M. high tide on December 15. Fishermen claim that tankers frequently "shoot the shoals" to save time and money. In this case, however, the captain's last estimate of his arrival time in Salem, as he reported to his agent in Boston, was 9 P.M., four hours after high tide. Thus he could not have been trying to make the afternoon tide, which he would have needed to dock at the discharge terminal. That alone dones not rule out the possibility of his taking a shortcut, but I have found no evidence that he was taking that risk when he ran aground. I believe the captain was looking for the lightship, as he has testified, and that he was lost.

R.W.